INTERNATIONAL
WILDLIFE
ENCYCLOPEDIA

THIRD EDITION

Volume 20
TRE–WAT

Marshall Cavendish Corporation
99 White Plains Road
Tarrytown, New York 10591–9001

Website: www.marshallcavendish.com

Library of Congress Cataloging-in-Publication Data

Burton, Maurice, 1898-
    International wildlife encyclopedia / [Maurice Burton, Robert Burton] .-- 3rd ed.
        p. cm.
    Includes bibliographical references (p.).
    Contents: v. 1. Aardvark - barnacle goose -- v. 2. Barn owl - brow-antlered deer -- v. 3. Brown bear - cheetah -- v. 4. Chickaree - crabs -- v. 5. Crab spider - ducks and geese -- v. 6. Dugong - flounder -- v. 7. Flowerpecker - golden mole -- v. 8. Golden oriole - hartebeest -- v. 9. Harvesting ant - jackal -- v. 10. Jackdaw - lemur -- v. 11. Leopard - marten -- v. 12. Martial eagle - needlefish -- v. 13. Newt - paradise fish -- v. 14. Paradoxical frog - poorwill -- v. 15. Porbeagle - rice rat -- v. 16. Rifleman - sea slug -- v. 17. Sea snake - sole -- v. 18. Solenodon - swan -- v. 19. Sweetfish - tree snake -- v. 20. Tree squirrel - water spider -- v. 21. Water vole - zorille -- v. 22. Index volume.
    ISBN 0-7614-7266-5 (set) -- ISBN 0-7614-7267-3 (v. 1) -- ISBN 0-7614-7268-1 (v. 2) -- ISBN 0-7614-7269-X (v. 3) -- ISBN 0-7614-7270-3 (v. 4) -- ISBN 0-7614-7271-1 (v. 5) -- ISBN 0-7614-7272-X (v. 6) -- ISBN 0-7614-7273-8 (v. 7) -- ISBN 0-7614-7274-6 (v. 8) -- ISBN 0-7614-7275-4 (v. 9) -- ISBN 0-7614-7276-2 (v. 10) -- ISBN 0-7614-7277-0 (v. 11) -- ISBN 0-7614-7278-9 (v. 12) -- ISBN 0-7614-7279-7 (v. 13) -- ISBN 0-7614-7280-0 (v. 14) -- ISBN 0-7614-7281-9 (v. 15) -- ISBN 0-7614-7282-7 (v. 16) -- ISBN 0-7614-7283-5 (v. 17) -- ISBN 0-7614-7284-3 (v. 18) -- ISBN 0-7614-7285-1 (v. 19) -- ISBN 0-7614-7286-X (v. 20) -- ISBN 0-7614-7287-8 (v. 21) -- ISBN 0-7614-7288-6 (v. 22)
    1. Zoology -- Dictionaries.   I. Burton, Robert, 1941-   . II. Title.

    QL9 .B796 2002
    590'.3--dc21

                                                    2001017458

Printed in Malaysia
Bound in the United States of America

07 06 05 04 03 02 01   8 7 6 5 4 3 2 1

Brown Partworks
Project editor: Ben Hoare
Associate editors: Lesley Campbell-Wright, Rob Dimery, Robert Houston, Jane Lanigan, Sally McFall, Chris Marshall, Paul Thompson, Matthew D. S. Turner
Managing editor: Tim Cooke
Designer: Paul Griffin
Maps: Dax Fullbrook, Seth Grimbly
Picture researchers: Brenda Clynch, Becky Cox
Illustrators: Ian Lycett, Catherine Ward
Indexer: Kay Ollerenshaw

Marshall Cavendish Corporation
Editorial director: Paul Bernabeo

Authors and Consultants

Dr. Roger Avery, BSc, PhD (University of Bristol)

Rob Cave, BA (University of Plymouth)

Fergus Collins, BA (University of Liverpool)

Dr. Julia J. Day, BSc (University of Bristol), PhD (University of London)

Tom Day, BA, MA (University of Cambridge), MSc (University of Southampton)

Bridget Giles, BA (University of London)

Leon Gray, BSc (University of London)

Tim Harris, BSc (University of Reading)

Richard Hoey, BSc, MPhil (University of Manchester), MSc (University of London)

Dr. Terry J. Holt, BSc, PhD (University of Liverpool)

Dr. Robert D. Houston, BA, MA (University of Oxford), PhD (University of Bristol)

Steve Hurley, BSc (University of London), MRes (University of York)

Tom Jackson, BSc (University of Bristol)

E. Vicky Jenkins, BSc (University of Edinburgh), MSc (University of Aberdeen)

Dr. Jamie McDonald, BSc (University of York), PhD (University of Birmingham)

Dr. Robbie A. McDonald, BSc (University of St. Andrews), PhD (University of Bristol)

Dr. James W. R. Martin, BSc (University of Leeds), PhD (University of Bristol)

Dr. Tabetha Newman, BSc, PhD (University of Bristol)

Dr. J. Pimenta, BSc (University of London), PhD (University of Bristol)

Dr. Kieren Pitts, BSc, MSc (University of Exeter), PhD (University of Bristol)

Dr. Stephen J. Rossiter, BSc (University of Sussex), PhD (University of Bristol)

Dr. Sugoto Roy, PhD (University of Bristol)

Dr. Adrian Seymour, BSc, PhD (University of Bristol)

Dr. Salma H. A. Shalla, BSc, MSc, PhD (Suez Canal University, Egypt)

Dr. S. Stefanni, PhD (University of Bristol)

Steve Swaby, BA (University of Exeter)

Matthew D. S. Turner, BA (University of Loughborough), FZSL (Fellow of the Zoological Society of London)

Alastair Ward, BSc (University of Glasgow), MRes (University of York)

Dr. Michael J. Weedon, BSc, MSc, PhD (University of Bristol)

Alwyne Wheeler, former Head of the Fish Section, Natural History Museum, London

Picture Credits
Heather Angel: 2767, 2768, 2770, 2870, 2874; **Ardea London Ltd:** John Mason 2806, Adrian Warren 2809; **Neil Bowman:** 2793, 2800, 2845, 2859; **Bruce Coleman:** Trevor Barrett 2780, Erwin & Peggy Bauer 2855, 2862, Mr. J. Brackenbury 2824, Bruce Coleman Inc 2851, Fred Bruemmer 2796, Bob & Clara Calhoun 2746, John Cancalosi 2750, Robert P. Carr 2788, Gerald S. Cubitt 2764, Jeff Foott 2787, Bob Glover 2858, Tore Hagman 2856, HPH Photography 2867, 2799, 2741, Wayne Lankinen 2802, Werner Layer 2749, Robert Maier 2784, 2849, Luiz Claudio Marigo 2757, 2803, Rita Meyer 2781, Marie Read 2745, 2833, Hans Reinhard 2751, John Shaw 2860, Kim Taylor 2752, 2760, 2761, 2861, Colin Varndell 2857, Jim Watt 2797, Staffan Widstrand 2853, Rod Williams 2841, 2846, Gunter Ziesler 2835; **Corbis:** Dewitt Jones 2828, Jeffrey L. Rotman 2819, Uwe Walz 2844, Lawson Wood 2783, **Chris Gomersall:** 2748, 2843, 2791, 2792; **Natural Visions:** Brian Rogers 2801; **NHPA:** A.N.T. 2762, 2817, 2818, Daryl Balfour 2866, A.P. Barnes 2810, G.I. Bernard 2812, 2816, 2825, 2871, N.A. Callow 2826, Laurie Campbell 2742, 2794, 2837, Stephen Dalton 2838, 2872, 2873, 2876, 2877, Nigel J. Dennis 2798, 2814, Pavel German 2848, 2868, Melvin Grey 2823, Ken Griffiths 2763, Martin Harvey 2847, Daniel Heuclin 2813, Hellio & Van Ingen 2779, E.A. Janes 2789, B. Jones & M. Shimlock 2743, 2758, T. Kitchin & V. Hurst 2865, Stephen Krasemann 2795, Gerard Lacz 2839, Yves Lanceau 2782, Mike Lane 2790, Lutra 2753, Trevor McDonald 2744, 2759, 2820, David Middleton 2832, Haroldo Palo Jr. 2756, 2808, William Paton 2840, Otto Pfister 2807, Rod Planck 2854, Jany Sauvanet 2804, Kevin Schafer 2836, 2852, John Shaw 2785, Eric Soder 2850, Robert Thompson 2811, Roger Tidman 2831, Michael Tweedie 2805, 2864, Alan Williams 2740, 2777, Norbert Wu 2775, 2830; **Oxford Scientific Films:** Kathie Atkinson 2869, G.I. Bernard 2863, M.A. Chappell 2821, J.A.L. Cooke 2769, David M. Dennis 2815, Jack Dermid 2822, Z. Leszczynski 2875, Norbert Wu 2829; **Still Pictures:** J.J. Alcalay 2776, John Cancalosi 2786, M. & C. Denis-Hout 2842, Juan Carlos Munoz 2834, Roland Seitre 2778, Norbert Wu, 2773, 2774, Gunter Ziesler 2807. **Artwork:** Catherine Ward 2754, 2765, 2771, 2827.

Contents

TREE SQUIRREL

Squirrels are sometimes nicknamed tree rats because of their rodentlike appearance. Pictured is a gray squirrel.

MANY SQUIRRELS have a primarily arboreal (tree-living) lifestyle. Tree squirrels are classified as the 28 species belonging to the genus *Sciurus* that are distributed over Europe, most of Asia and North America from southern Canada to the eastern United States, although scientists currently subdivide this genus into seven subgenera. Two of the best-known and most widespread squirrels, which are discussed here, are the gray squirrel, *S. carolinensis*, of North America and the red squirrel, *S. vulgaris*, of Europe and Asia.

Tree squirrels differ from ground squirrels and flying squirrels (both discussed elsewhere) in having bushy tails, 8–12 inches (20–31 cm) long, similar to the length of the head and body combined. They range from 16–24½ inches (40–63 cm) in total length and weigh between 7–35 ounces (200–1,000 g). Their coat varies in color but can be gray or red, sometimes black, with white or cream, sometimes yellow or orange, underparts. The winter coat is usually slightly different from the summer coat, and although the body fur is molted twice a year, the tail is molted only once. The tail is always well furred, highly tufted and fluffy. In many species it is flattened and feathery rather than bushy. The ears sometimes have tufts of hair at the tips but these tufts are retained for only part of the year in some species. The four toes on the front feet and the five toes on the hind feet bear sharp claws that are useful for climbing trees.

Arboreal acrobats

Tree squirrels forage on the ground but quickly escape to trees when they are disturbed. They move rapidly up a trunk, their first leap taking them 3–4 feet (90–120 cm) up, after which their sharp claws are used in a bounding climb. They run along the branches or hang from them, traveling upside down by using all four feet, one over the other. The gray squirrel in particular is a skilled acrobat. It hangs by its hind feet from one branch to reach food on a branch below. It also leaps gracefully from the outer branches of one tree to those of another over a distance of 12 feet (3.6 m). As it sails through the air, the legs are spread-eagled and the flattened tail acts as both a balancer and a rudder.

A tree squirrel's usual reaction to an intruder is to chatter at it, or scold, and then to disappear behind a trunk or stout branch, all the time keeping the trunk or branch between itself and its predator. When forced to do so, it may drop to the ground for escape, from heights of 30 feet (9 m) or more, plunging directly into undergrowth or dropping straight down on all fours onto bare earth or short turf.

Scattered stores

Although primarily vegetarian, eating nuts, berries, soft fruits, buds and some fungi, most tree squirrels take birds' eggs and nestlings, even carrion. They traditionally are hoarders but so far as the gray and the European red squirrels are concerned, they seldom cache food in the way traditionally attributed to them. Stores of nuts buried in the ground or in a hollow tree are more likely to be the work of field mice. A squirrel buries nuts, acorns or berries singly and well

TREE SQUIRREL

CLASS	**Mammalia**
ORDER	**Rodentia**
FAMILY	**Sciuridae**
GENUS	***Sciurus*; 7 subgenera**
SPECIES	**28, including red squirrel, *S. vulgaris*; gray squirrel, *S. carolinensis***

WEIGHT
7–35 oz. (200–1,000 g)

LENGTH
Head and body: 8–12½ in. (20–32 cm); tail: 8–12 in. (20–31 cm)

DISTINCTIVE FEATURES
Small, rodentlike animal; highly tufted and fluffy tail, as long as body; varied color, red or gray, sometimes black; underparts lighter than upperparts and tail

DIET
Nuts, seeds, young tree shoots, buds and fruit; sometimes birds' eggs, insects and mushrooms

BREEDING
Age at first breeding: 1 (female), 2 (male); breeding season: year-round in Northern Hemisphere, with peaks in winter and late spring; number of young: 4 to 10; gestation period: 40 days; breeding interval: 6 months

LIFE SPAN
Up to 5 years; up to 24 years in captivity

HABITAT
All types of forests and woodland in both humid and arid environments

DISTRIBUTION
Europe, Asia and eastern North America

STATUS
Red squirrel: near-threatened; several subspecies of tree squirrels threatened or endangered in North America

Tree squirrel ▢ Gray ▢ Red

Female tree squirrels can breed in their first year, but males are not successful breeders until their second year. Above is a South African tree squirrel, **Paraxerus capapi.**

spaced out. It carries the food in its mouth, stops at a chosen spot, digs a hole with a quick action of the forefeet, just deep enough to take the nut or acorn, and then buries it by pushing back the earth with the forefeet.

Importance of chewing

Zoologists have discovered that a young gray squirrel does not open a nut efficiently the first time but learns to gnaw it open. Although all adult gray squirrels open nuts in the same way, showing that the pattern of this behavior is species-specific (members of the species instinctively perform the action in the same way) the instinct has to be reinforced by learning.

The front teeth of squirrels grow continuously at the roots, and common belief holds that their habit of chewing objects other than food is necessary to keep the teeth worn down to normal length. Some of the damage done to trees comes from this chewing habit, but the way gray squirrels chew the metal labels on ornamental trees in parks and gardens is perhaps the best illustration of the habit. Recently, however, zoologists have discovered that squirrels, as well as rats, constantly grind their teeth when not otherwise engaged, and it is this grinding action that keeps the teeth at the required length.

The red squirrel,
S. vulgaris, *now classed*
as near threatened,
is in danger of being
out-competed by the
introduction of the
larger gray squirrel in
some areas of the red
squirrel's range.

Nest of twigs and leaves

Tree squirrels build nests in the branches of trees, and use them for several purposes, such as to serve as a nursery. Each squirrel usually builds several nests in adjacent trees. They are bulky structures composed of twigs, strips of thin bark, leaves and moss. Some are cup-shaped, whereas others are domed. Sometimes a squirrel may adapt an old bird's nest, such as one made by a crow, to use as a foundation. The breeding or nursery nest often is a huge ball of leaves and sticks in a roomy hollow in a tree trunk. The breeding season of the European red squirrel varies with latitude. In the northern parts of the range the only breeding period is spring. In the southern parts there are two periods, January to April and May to August. Gestation is 40 days, the litters consisting of 4 to 10 young, usually 5 to 7, born blind and naked. They are weaned at 7 to 10 weeks. In the gray squirrel mating can occur year-round, but usually takes place between early January and August. There are usually two litters a year. Gestation lasts 44 days, and other details of development in the gray squirrel are much the same as those of the red squirrel.

Declining predators

Tree squirrels are remarkably skillful not only in moving among trees but also in keeping out of sight once they are alerted to possible danger. In much of Europe the natural predators of squirrels have been largely eliminated. What kinds of animals these predators could be throughout the range of tree squirrels can be judged from the natural predators of gray squirrels in North America. They include the goshawk, red-shouldered hawk, barred owl, horned owl and tree-climbing snake. On the ground, squirrel predators include the fox, the bobcat and finally, and probably most importantly, the pine marten, which can move through trees at least as skillfully as the nimble squirrel.

Rapid tooth growth

Rodents' incisors grow at a remarkable speed. If they fail to bite on and grind away at each other, the teeth curve into the mouth and eventually lock the jaws so that the animal dies of starvation. Zoologists are now able to make accurate measurements of the rate of incisor growth. Although the figures for tree squirrels are not known, they probably compare fairly closely with those that are known for other, similar species. The incisors of the common rat, for example, grow up to 6 inches (15 cm) in a year, those of the guinea pig up to 10 inches (25 cm) and those of the pocket gopher, which uses its teeth for digging and therefore needs a rapid growth rate to counteract a high rate of wear, grow up to 14 inches (35 cm) in a year.

TRIGGERFISH

THE TRIGGERFISH'S BODY is deep and compressed and, seen from the side, it is almost diamond shaped. The head occupies about a third of the length, the mouth is small and the eyes are fairly large. The triggerfish has three spines on the back, the third of which can be minute. One of these spines has a locking device that ensures it remains erect until the spine behind it is depressed to release it. Just behind the eye is a short dorsal fin with spines, the first two making the locking device. The second dorsal fin is long and high; the anal fin is the same shape and lies exactly opposite it. The pectoral fins are small and the pelvic fins are no more than short spines. The body, seldom more than 2 feet (60 cm) long, is covered with small, rhomboidal bony plates, their outer surfaces bearing one or more small spines. Those triggerfish in the family Balistidae are often boldly marked and patterned.

The filefish of the family Monachanthidae, comprising 31 genera and 95 species, are close relatives of the triggerfish. They are similar in shape but usually grow to only 1 foot (30 cm) long, and usually have two dorsal spines, which can be locked in place, level with or in front of the eye. The second spine is usually smaller than the first, and may be absent. The filefish scales carry more spines, so its surface is rough, somewhat in the manner of a file. The second dorsal fin and the anal fin are not as high. In place of pelvic fins there is a spine on the pelvic bone, which can move freely and is connected to the body by a wide flap of skin. Filefish and triggerfish live mainly in tropical coastal waters.

Swimming techniques
Triggerfish and filefish have quite different swimming techniques. Although both swim slowly with the body rigid, triggerfish swim by simultaneously flapping the second dorsal and anal fins. Filefish are driven through the water by waves passing backward along these fins.

Another contrast between the fish is evident in their means of defense. The triggerfish seems to use its locking spine when it takes refuge in a crevice in rock or coral. It erects the spine and cannot be pulled out. It is impossible to press the spine down with the finger but the fish can lower it by dropping the second spine, which releases the large spine in front of it. The fish can be made to lower the first spine if the third one is pressed down. When a triggerfish is seen from any but the side view, its compressed body presents only a fairly thin edge, helping the fish to

conceal itself. According to United States marine biologist William Beebe, at least one species of filefish, the scrawled fish, *Aleuterus scriptus*, has an equally remarkable means of defense. It stands on its head among clumps of eelgrass, its fins gently waving. With its mottled green color the fish is hard to distinguish from the eelgrass.

Shell crackers
Both triggerfish and filefish have teeth implanted in sockets in the jaws, which is unusual for any fish. Triggerfish usually have seven chisel-like teeth in each jaw, which they use to crush holes in the shells of mussels, oysters and clams to eat the soft flesh inside, or to crack crabs and other crustaceans. They also eat carrion. Scientists believe filefish are vegetarian. The five teeth the fish have in each jaw are developed for nibbling.

Spiny deterrent?
There is still much scientific work to be done regarding the breeding habits and the predators of triggerfish. It is likely that the spines stick in the throat or damage the mouth of a predator. Their striking colors may well act as a warning, and any predator taking a bite and suffering injury is likely to leave the next triggerfish or filefish alone. It is also possible that both types of the fish may be poisonous, but so far the only evidence of this is that triggerfish may sometimes be poisonous to eat when they themselves

The Indonesian clown triggerfish, Balistoides conspicillum, is so-named because of its bright and colorful appearance.

Triggerfish have sharp teeth, with which they crush invertebrates' shells. This yellow-margined triggerfish, Pseudobalistes flavimarginatus, *is trying to open a mollusk.*

TRIGGERFISH

CLASS	**Actinopterygii**
ORDER	**Tetraodontiformes**
FAMILY	**Balistidae**
GENUS	**11**
SPECIES	**36, including clown triggerfish,** ***Balistoides conspicillum;*** **gray triggerfish,** ***Balistes carolinensis*** **(detailed below)**

WEIGHT
3 lb. (1.3 kg)

LENGTH
24 in. (60 cm)

DISTINCTIVE FEATURES
Deep, diamond-shaped body, flattened laterally; small mouth; large pointed teeth; 2 large spines behind gill slit; 2 dorsal fins, the first with 3 spines; green, gray or brown

DIET
Bottom-dwelling mollusks and crustaceans

BREEDING
Breeding season: summer, in Mediterranean; male guards eggs; hatching period: 2 days

LIFE SPAN
Not known

HABITAT
Bays, harbors, lagoons and seaward reefs

DISTRIBUTION
North Atlantic and Mediterranean south to Angola; Novia Scotia, Bermuda and northern Gulf of Mexico south to Argentina

STATUS
Not listed

have eaten contaminated shellfish or carrion. None of the above explains the function of the pelvic spine, especially that of the filefish. It has been suggested that the fish uses this spine to fix itself in crevices in rocks or coral, as a defensive measure, just as the triggerfish uses its spines.

Triggerfish and filefish may not be as defenseless as their slow-moving habits suggest. They can inflict deep bites with their strong, sharp teeth and also make sounds, usually a sign of strong territorial instinct or an indication of an animal's ability to defend itself, or both. Some triggerfish grate their teeth together in the roof and floor of the throat. Other types, as well as the filefish, rub together the bases of their fin spines. In all the fish, the sounds are amplified by the swim bladder. The black triggerfish, *Melichthys niger*, of the Caribbean, makes a rapid puttering sound by vibrating a membrane in front of its pectoral fins.

After mating, the triggerfish digs a pit or nest in the sand or coral rubble 8 feet (2.4 m) in diameter and 2 feet (60 cm) deep, in which eggs are laid and glued to the bottom; they are known as demersal eggs, meaning they are related to the sea bottom. Where it is impossible to glue the eggs to the sand, the fish carries bits of coral rocks to the nest and drops them onto the eggs. The female guards the nest aggressively, but the male stands guard less frequently.

Colorful creatures

The gray triggerfish, *Balistes carolinensis,* living in the tropical eastern Atlantic and the Mediterranean, often wanders as far north as the waters around Britain in summer. It is grayish brown to greenish brown with a violet tinge on the back and blue bands and yellow or black spots on the dorsal and anal fins, a sober coloring compared with most triggerfish species. The orange-lined triggerfish, *Balistapus undulatus,* of the Indian

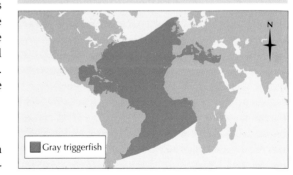

Gray triggerfish

Ocean and South Pacific is purplish with small orange spots on the face, orange lines all over the body and golden orange fins. The blackbar triggerfish, *Rhinecanthus aculeatus,* or *humahumanuku-nuka-a-puaa* of the southern Pacific also has a very distinct pattern.

TROGON

THE TROGONS ARE highly colored, fairly large, tropical birds, the most brilliant of which, the quetzal or resplendent trogon, is described elsewhere. Although ranging throughout the Tropics, the 39 or more species of trogons are remarkably alike in body form. They range in size from 9–13 inches (22.5–32.5 cm) and have rounded wings and short blunt bills. The tail is long and, although the tail feathers are of unequal length, it is kept folded, so it appears to be square. The feet are very small, with two toes facing forward and two facing backward. Trogons have the second, or inner, toe facing backward instead of the fourth, or outer, toe as in woodpeckers, parrots and some other birds. Trogons also have delicate skin and soft feathers. The male's plumage is brighter than the female's. The upperparts are usually green, blue or brown, and the underparts are bright red, orange or yellow.

Trogons range from the United States, where most are found, through Africa south of the Sahara, India, southern China, the East Indies and the Philippines. Very few are found outside the Tropics. The elegant trogon, *Trogon elegans*, breeds as far north as northern Arizona, while the Narina trogon, *Apaloderma narina*, ranges from Liberia to the Cape Province.

Hidden away

Trogons are mainly forest birds but are also found in those plantations where there are sufficient trees. As with so many brightly colored birds of the tropical forests, trogons are quite inconspicuous because the bold patches of color break up the outlines of their bodies in the light and shadow of the upper layers of the forest. In addition, trogons are commonly described as lethargic because they perch motionless, with their bodies upright and their long tails hanging straight down, often for minutes at a time. Perched trogons often have a hump-backed appearance.

Their simple calls consist of various coos, hoots and whistles that serve as communication between nesting pairs, alarm calls or when the male announces his territory. The calls are sometimes ventriloquial, which further increases the difficulty of locating the trogons.

Plucking insects and fruit

Trogons are most easily spotted when they are feeding because they forage like flycatchers in the upper and outer canopy of trees. Flying insects are only occasionally caught, however, and the most frequent prey of trogons is nonflying insects, which they pluck from leaves and twigs while hovering.

The broad bill is surrounded by bristles that form a net for trapping insects. Other small animals such as snails, small tree frogs and lizards are also plucked from the foliage.

Many trogons eat fruit by the same method, plucking it from the stem in flight and then carrying it back to the perch to eat it. The trogons

Male slaty-tailed trogon, T. massena, of Panama. Trogons usually have an undulating flight, but only montane forest species fly more than short distances.

Male elegant or coppery-tailed trogons have a conspicuous bright red breast and white breast band with gray wings.

TROGON

CLASS	**Aves**
ORDER	**Trogoniformes**
FAMILY	**Trogonidae**

GENUS AND SPECIES **39 or more species, including elegant trogon, *Trogon elegans* (detailed below)**

WEIGHT
2½ oz. (70 g)

LENGTH
11–12 in. (28–30.5 cm)

DISTINCTIVE FEATURES
Male: dark green upperparts, long rufous tail, stout yellow bill, orange eye ring, white breast band and conspicuous scarlet belly. Female: similar structurally, light brown upperparts, rufous tail, white eye ring and scarlet vent area.

DIET
Insects, small lizards, frogs, snails; small fruits

BREEDING
Age at first breeding: 1 year; breeding season: May–June; number of eggs: 2 to 3; incubation period: 17–19 days; fledging period: 14–30 days; breeding interval: 1 year

LIFE SPAN
Not known

HABITAT
Arid to semiarid woodland and thorn forest

DISTRIBUTION
Arizona in the United States south to northwestern Costa Rica in South America

STATUS
Fairly common

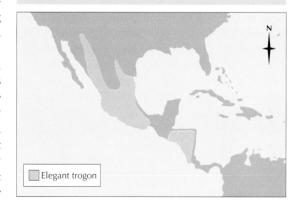

Elegant trogon

found in the United States and South America are particularly fond of fruit. Their notched bills probably assist in the rapid plucking necessary for the successful execution of this aerobatic act.

Plaintive calls

Trogons are usually solitary or found in pairs, but several birds may gather at fruiting trees. During the breeding season, the males of some species gather in small groups. In August, families of elegant trogons are often seen, including males, females and immature offspring, all feeding or perched together in the same area.

Most trogons from the United States and South America have a weak courtship song; it is little more than an elaborate form of the simple call. Trogon songs are typically repeated hollow or plaintive hoots. For example, the mountain trogon of Mexico, *T. mexicanus*, has a song that consists of plaintive, paired whistles. It usually sings three pairs of whistles per series, but sometimes sings 15 or more pairs of notes, for example, *kyow-kyow... kyow-kyow...* and so on. The calls also include clucks and chattering, often produced by the males to attract females to their territories. Most trogons are heard far more often than they are seen.

Cavity nests

The nest is always in a cavity, such as the abandoned hole of a woodpecker or a natural hole in a rotten stump. Very often a special hole is excavated, but as trogons' bills are short and weak they can dig in only very rotten wood. Some trogons, however, make their homes in the nests of termites or wasps. For example, the gartered trogon, *T. violaceus*, uses the papery wasps' nests that are suspended from the branches of trees. Before starting to dig, the bird eats the wasps, and as it tears out the combs, it also eats the larvae.

The nest of the blue-throated trogon, *T. rufus*, is no more than a pear-shaped cavity, 3–4 inches (7.5–10 cm) across and twice as high, in which the trogon sits with its long tail turned up over its back. The nest is not lined and the two to four white, oval eggs are incubated for about 18 days. Both members of the pair share the work of digging the hole, incubating and feeding the chicks. In the blue-throated trogon the male incubates the eggs by day and the female from late afternoon to the following morning. When they first come off the nest, young elegant trogons are dappled with white and do not have a long tail.

Instinctive feeding

One observer has described watching a male mountain trogon that was rather slow in feeding its chicks. For the first day or two after they had hatched the male brought insects to the nest but did not give them to the chicks. Instead, he ignored their clamoring and flew away. This behavior contrasts with the enthusiasm of the male song tanagers (described elsewhere), which sometimes try to feed their eggs. A bird must start feeding its offspring instinctively, because there is no previous training. The instinct is presumably prompted by the sight of the chicks in the nest.

In some patterns of instinctive behavior there is more than a simple automatic switching-on in response to a stimulus; the actions have to be perfected by learning. In the case of the mountain trogon described above, the feeding behavior was switched on but there must have been some type of temporary biological fault that took a day or two to clear. After this time the trogon was able to complete the necessary sequence of actions by dropping the food into the chicks' open mouths.

*This male white-tailed trogon, **T. viridis**, inhabits the Utria National Park, western Colombia.*

TROPIC BIRD

A white-tailed tropic bird from the Seychelles in the Indian Ocean. Tropic birds spend the majority of their time in flight above the oceans.

TROPIC BIRDS ARE seabirds with an easy wheeling flight, their graceful appearance enhanced by their long tail plumes. They are related to cormorants, pelicans and boobies, although in appearance and habits they resemble the terns in some respects and the petrels in others.

Tropic birds range from 31–40 inches (78–100 cm) long, but nearly half this length is made up of two very long, streaming tail feathers. Without these feathers, tropic birds are only 14–19 inches (35–47.5 cm) long. The bill is thick and slightly curved, the legs are short and the feet are webbed. The plumage is white, sometimes tinged with pink, with black eye stripes and black on the wings.

There are three species of tropic birds, all living in the Tropics. English explorer and colonist John Smith (c.1580–1631), writing in 1622, stated that the name tropic bird derived from the places where the bird was most often seen. The red-billed tropic bird, *Phaethon aethereus*, is found in the Atlantic, Pacific and Indian Oceans, including the Red Sea and the Caribbean Sea. The upperparts of this species are barred with black. The white-tailed tropic bird, *P. lepturus*, also lives in these three oceans and is common in the Caribbean Sea. It can be distinguished by the solid black marking on the wings. The largest species, the red-tailed tropic bird, *P. rubricauda*, is almost pure white or pink except for the black eye stripes and the red tail streamers. It is found in the South Pacific and the southern Indian Ocean, and breeds on the islands of Hawaii and on islands in the Banda Sea, Indonesia, and many in the Indian Ocean.

At home in the air

Tropic birds have a fast, rather pigeonlike flight, and they spend more time in the air than any of their relatives except perhaps the frigatebirds (discussed elsewhere). Outside the breeding season they may be found in ones and twos well out to sea. Tropic birds do not settle on the sea very frequently, and when they do, they are poor swimmers. They come to land only during the breeding season.

RED-TAILED TROPIC BIRD

CLASS **Aves**

ORDER **Pelecaniformes**

FAMILY **Phaethontidae**

GENUS AND SPECIES **Phaethon rubricauda**

WEIGHT
21–29 oz. (600–835 g)

LENGTH
Head to tail: 31–32 in. (78–81 cm), including 12–14-in. (30–35-cm) tail streamers

DISTINCTIVE FEATURES
Predominantly white; black eye stripe; bright red bill; longish wings; wirelike red tail streamers, often difficult to see

DIET
Mostly fish, especially flying fish; squid and crustaceans

BREEDING
Age at first breeding: not known; breeding season: year-round in some places; number of eggs: 1; incubation period: 42–46 days; fledging period: 67–91 days; breeding interval: 1 year

LIFE SPAN
9 years

HABITAT
Tropical and subtropical seas; breeds in inaccessible cliff areas on small, remote oceanic islands

DISTRIBUTION
Breeds in Hawaii and 9 island groups in South Pacific; Banda Sea, Indonesia; also, many islands in Indian Ocean

STATUS
Not globally threatened

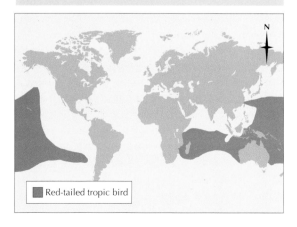

Red-tailed tropic bird

The legs of tropic birds are so short and set so far back on their bodies that the birds can only shuffle awkwardly and can hardly lift their bodies off the ground. Their nesting places are usually inaccessible places on cliff faces, from which they can launch themselves into the air with the minimum of effort.

Tropic birds, such as this red-billed tropic bird, come to land only during the breeding season, when a single egg is laid in the nest.

Plummet to feed

Tropic birds closely resemble terns in their feeding habits. They fly at a height of about 50 feet (15 m), with the head turned down and the bill held vertical in a manner reminiscent of terns. When they spot food, tropic birds hover on rapidly beating wings and then plunge into the sea with hardly a splash. A moment later they surface, shake themselves while holding their wings and tail raised and then spring into the air. The birds eat fish, squid and crustaceans, which they grasp with the sharp edges of the bill.

Petrel-like behavior

On the Galapagos and Ascension Islands, and a few other places, tropic birds nest year-round, probably because there is a steady supply of food throughout the year. Nesting takes place on islands, starting with a long courtship period in which small groups of tropic birds chase each other through the sky.

Courtship also takes place on the nests, which are no more than bare patches of ground in crevices or under rocks. There is competition for suitable sites, and fighting often breaks out, the birds sparring with their bills and sometimes injuring each other.

Hunting high above the surface of the ocean, tropic birds hover over their prey of fish or squid before diving into the sea to claim it. Pictured is a red-billed tropic bird.

A single brown-and-white egg is incubated by both parents for periods of several days at a time. The chick emerges in about 6 weeks, and another 2 months or more elapse before it finally leaves the nest. It hatches with a covering of down and develops slowly, being left alone after a while as both parents go away to feed. During this stage many tropic bird chicks are killed in the squabbles for nest sites. The chick leaves the nest when it is 2 months old. By this time the parents have deserted it and it has lost weight. This desertion, the long nestling period and the bird's shuffling gait on land are all reminiscent of the behavior of the petrels.

Cropping their tails

Tropic birds nest on islands, sometimes well out to sea, and consequently they suffer from few predators. The main cause of death seems to be the attacks by adults on nestlings. Like many other seabirds that nest in isolated places, tropic birds are fearless and even allow themselves to be lifted off the nest.

In the Caribbean this tameness has led to their being killed by rats, and in Bermuda their eggs were once collected for food. Otherwise, tropic birds have suffered little from the actions of humans. However on some Pacific islands the local peoples pulled the long tail feathers from nesting tropic birds for use as ornaments. Perhaps the main danger today comes from cats or rats introduced onto islands.

While a parent is on the nest, it not only sits tight, but also tries to defend itself in the face of an attack. This lays it open to a quick death from a cat or a full-grown rat. When the parent tropic birds leave the nests, the chicks become even more vulnerable to these two predators.

Boatswain birds

Tropic birds are a familiar sight to sailors in tropical regions. They do not follow ships in the manner of albatrosses or storm petrels, but do investigate ships that cross their path. In French, tropic birds are known as *paille-en-queue*, and English sailors once called them marlin-spikes, based on the outline of their broad bodies and their long, narrow tail feathers. A more common alternative is bo'sun bird or boatswain bird, which may have been prompted by the resemblance of the twittering calls of the red-billed tropic bird to the boatswain's pipe. Others claim the name comes from the birds' rolling gait on land. The generic name of tropic birds derives from Greek myth: Phaethon, the son of Apollo, plummeted headlong into the sea from the sky.

TROUT

THERE ARE 12 GENERA and 183 species of trout, including 21 subspecies. The European or brown trout, *Salmo trutta*, for example, is divided into six subspecies. The Aral trout, *S. t. aralensis*, is endemic to the Aral Sea and is found in Central Asia. The brown trout, *S. t. fario*, is a North Atlantic species ranging from Norway to southern France and Greece. Widely introduced to North and South America, the sea trout, *S. t. trutta*, is native to Europe and Asia. The other subspecies are the lake trout, *S. t. lacustris*, which is widespread across Europe, *S. t. macrostigma*, which is found only in the island of Corsica in the Mediterranean and the Amn-Darya trout, *S. t. oxianus*, which is native to the former Soviet Union.

The European brown trout and lake trout are greenish brown, the flanks being lighter than the back, and the belly yellowish. They are covered with many red and black spots, the latter being surrounded by pale rings. There are also spots on the gill covers. These two trout and the sea trout resemble the salmon in both shape and appearance, except that the angle of the jaw extends to well behind the eye and the adipose (fatty) fin is tinged with orange.

The North American species are similar. The cut-throat trout, *Oncorhynchus clarki*, is so-named for the two red marks across its throat. There are three subspecies of this fish: the cut-throat trout of the eastern Pacific, *O. c. clarki*, *O. c. lewisi* from North America and *O. c. pleuritians* from the Green and Platte Rivers and the Yellowstone River, Montana, in the United States.

The Dolly Varden trout, *Salvelinus malma*, is named for its conspicuous red spots, colored like the cherry ribbons worn by a character created by the novelist Charles Dickens. There are three subspecies that range from the northwest Pacific, the Arctic and North Pacific drainages and from Hokkaido, Japan, respectively.

In the brook trout, *S. fontinalis*, originally from North America, the pattern is more mottled but it also features red spots on the flanks. The rainbow trout, *O. mykiss*, originally from the eastern Pacific, has a reddish band running along the flanks. The lake trout of North America, *S. namaycush*, lives in deep water to about 400 feet (120 m). The golden trout, *O. aguabonita*, lives in water 8,000 feet (2,400 m) or more above sea level. Like the lake trout, this fish also comes from North America.

Trout, such as this rainbow trout, prefer cold, well-aerated upland waters where there is cover in the form of submerged rocks, undercut banks and vegetation.

Clean-living fish

Trout grow best in clear, aerated waters, and although they are sometimes found in turbid surroundings, this occurs only when the surface layers are well supplied with oxygen. They are readily affected by silt; it may spoil their spawning sites, reduce their food supply or act directly on the fish themselves. Laboratory experiments have shown that particles in suspension in the water, at a level as low as 270 parts per million, abrade the gills or cause them to thicken. The rate of growth of trout varies in other ways as well, often to a remarkable extent, with the conditions of their surroundings. Temperature, for instance, is very important, and an example can be seen at the time they resume feeding after the winter fast. Normally, trout stop feeding in the fall and resume in spring, in about March when the water reaches a temperature of 36° F (2° C) or more. In a mild winter they may begin feeding in December and continue until the first cold snap of the following fall.

The rate of growth also varies from one river to another, and from river to sea. Trout living in small streams grow more slowly than those in large rivers, and those that live in large bodies of fresh water grow more slowly than those living

Aquatic insects and their larvae are important constituents of the rainbow trout's diet.

TROUT	
CLASS	**Actinopterygii**
ORDER	**Salmoniformes**
FAMILY	**Salmonidae**
GENUS	**12**
SPECIES	**183, including sea trout, *Salmo trutta trutta* (detailed below); rainbow trout, *Oncorhynchus mykiss*; lake trout, *S. t. lacustris*; and cut-throat trout, *O. clarki*; others**

WEIGHT
Up to 110 lb. (50 kg)

LENGTH
Up to 4½ ft. (1.4 m)

DISTINCTIVE FEATURES
Deep, flattened caudal peduncle (narrow part at rear of body); square-cut tail; upper jawbone extends well beyond eye level; very small scales; numerous teeth; in streams, brownish color, darker on back, silvery on sides, numerous dark spots and orange-edged adipose (fatty) fin

DIET
Aquatic and terrestrial insects, mollusks, crustaceans and smaller fish

BREEDING
Number of eggs: up to 1,000; hatching period: 4–7 weeks

LIFE SPAN
Up to 14 years

HABITAT
Cold, well-oxygenated upland waters

DISTRIBUTION
Europe and Asia; introduced widely to the Americas and Australia

STATUS
Low risk

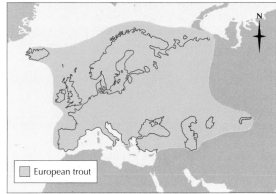

European trout

in the sea. A trout in a small river grows 2½, 5 and 8 inches (6.3, 12.5 and 20 cm) in its first, second and third years, respectively. Corresponding growth figures for the first three years of a sea trout are 3–5 inches (7.5–12.5 cm), 4–5 inches (10–12.5 cm) and 10–11 inches (25–27.5 cm).

A maturing diet

The diet of trout varies with their age. Fry eat mainly aquatic larvae of insects, rarely the adults. Later they eat large numbers of winged insects, as well as water fleas and freshwater shrimps. Adult trout eat mainly small fish as well as shrimps, insect larvae and adults, especially the winged insects. Sea trout feed on sprats, young herring and sand eels and also on a large percentage of small crustaceans, including shrimps and prawns.

Returning to spawn

Male trout begin to breed at 2 years, females at 3, returning to the place where they themselves were hatched. This homing instinct has been verified experimentally by scientists who transported marked trout to other parts of a river system and recorded them later back on their home ground. Breeding usually takes place from October to February, the time varying from one locality to another. Spawning normally occurs in running water, and the trout live in lakes going into the feeder streams.

For spawning the female makes a redd (nest) in gravelly shallows, digging a depression with flicks of her tail. As she lays her eggs, the male fertilizes them, stationing himself beside her but slightly to the rear. Scientists have discovered that the most successful redd is one with a current flowing downward through the gravel. The eggs hatch in about 40 days. The fry are ½–1 inch (1.3–2.5 cm) long at hatching, and the yolk sac is absorbed in 4 to 6 weeks.

Surrounded by predators

It is thought that about 94 percent of fry are lost during the first 3 to 4 months of their lives, after which the mortality drops to 20 percent. Eels are commonly thought to kill trout and to ravage the spawning grounds but there is no evidence of this. The chief predators of trout are water shrew, mink, the common rat and to some extent otters and herons. Another predator of small trout is larger trout. Well-grown specimens have sometimes been found to have another trout, 5–6 inches (12.5–15 cm) long, in their stomachs. In their cannibalism, trout vie with pike, traditionally regarded as predators, but which, with few exceptions, take only medium to large trout.

There are two other factors that contribute to trout depletion, apart from humans: competition for food from other animals, such as waterbirds and eels, and a lack of oxygen, especially during the winter. When the pools and lakes are frozen over, trout must rely on oxygen trapped under ice. This supply is replenished by oxygen given out by water plants. However, when the ice is blanketed by snow, light does not penetrate, the plants are not able to photosynthesize and the trout are asphyxiated.

Individual rainbow trout vary considerably in color, but each has a broad stripe running along each side and black spots on the sides, fins and back.

TROUT-PERCH

The trout-perch is widely distributed in North America, where it favors the backwaters of large, muddy rivers and lakes.

THE TWO SPECIES OF trout-perch live in North America. The blunt-nosed trout-perch, or sandroller, *Percopsis transmontana*, is found in the fresh waters of most of Canada, in Alaska and southward to Virginia, Kentucky, Missouri and Kansas. The second species, called simply the trout-perch, *P. omiscomaycus*, is more localized, being found only in the basin of the Columbia River in western North America. Both species are small, the sandroller measuring up to 6 inches (15 cm) long and the trout-perch reaching 8 inches (20 cm) in length.

Trout-perch have spotted bodies, a fairly pointed head and large eyes and when freshly caught have a distinctive translucent appearance. Although they resemble both trout (family Salmonidae) and perch (family Percidae) in certain respects, trout-perch are related to neither. Among the features they share with perch are their mouth shape and a body that is covered with spiny scales. Trout-perch also have spiny fins like those of a perch, the dorsal, anal and pelvic fins having one or more stout spines in their leading margin. In terms of number of fins, however, trout-perch are like trout. They also have the adipose (fatty) fin and naked head characteristic of all the Salmonidae.

An evolutionary offshoot

Trout-perch are common in the larger streams and in deep, clear lakes, especially those with sandy or gravelly bottoms. They spend the hours of darkness feeding in the shallows and move back into deeper waters during the day. Trout-perch are shoaling species. They eat aquatic insects, fish and small freshwater crustaceans, and are taken in turn by various kinds of predatory fish, including trout and pike (family Esocidae), and as live bait by fishers. At one time the two species were regarded as a kind of missing link between the salmon family and the perch family. However, more recent studies have shown that trout-perch are an evolutionary offshoot of the perch and were probably once much more widespread, with many more species.

In late May or early June in the southern parts of the range (slightly later in the north), trout-perch move to their spawning grounds.

TROUT-PERCH

CLASS	**Osteichthyes**
ORDER	**Percopsiformes**
FAMILY	**Percopsidae**
GENUS AND SPECIES	***Percopsis omiscomaycus***

LENGTH
8 in. (20 cm)

DISTINCTIVE FEATURES
Pale yellowish to silvery body color, frequently nearly transparent; about 10 dark spots in row along midline of back, 10 to 11 spots along lateral line, row of spots high on sides above lateral line; adipose fin; transparent fins transparent; small, weak spines in anal and dorsal fins; pectoral fins extend behind bases of pelvic fins; rough, ctenoid (toothed) scales

DIET
Insect larvae, amphipods and fish

BREEDING
Eggs and milt released in shallow water

LIFE SPAN
4 years

HABITAT
Lakes, rivers and deep-flowing pools of creeks, usually over sand

DISTRIBUTION
Western North America in basin of Columbia River

STATUS
Not threatened

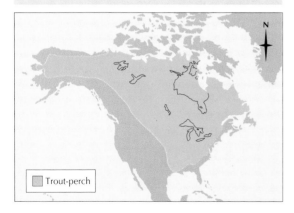

Trout-perch

They spawn in water not more than 3½ feet (106 cm) deep, so in the lakes the shoals move to sandbars, ascending feeder streams when no suitable shallow waters are available. In rivers, trout-perch swim upstream. There is much jostling among the members of the shoal as they sort themselves out, generally with one female attended by two or more males. Remaining close to the surface, the males press close to the female and release their milt as she releases her eggs. Fertilization takes place, and the eggs, 1.4 mm in diameter, sticky and heavier than water, sink to the bottom to adhere to the coarser gravel or rocks. Some populations of trout-perch spawn during the day, others exclusively at night. Trout-perch do not take care of the eggs once they have spawned nor do they look after the fry once they have hatched. Many adults die once spawning is over.

Piratical relative

Closely related to the trout-perch is the pirate-perch, *Aphredoderus sayanus*. It resembles the trout-perch, although it has a deeper body, a more square-ended tail and no adipose fin. Measuring up to 5 inches (13 cm) in length, the pirate-perch has olive-green to brown upperparts, is yellowish brown on the underside and has dark spots and blotches, usually in rows along the body. The pirate-perch lives in the eastern United States, from New York to Texas, in streams and standing waters. It is said to be very quarrelsome, both with fish of other species and with members of its own. This aggression stems largely from the pirate-perch's strongly territorial nature. The pirate-perch prefers waters with muddy beds, where there are debris and rotting leaves under which it can hide, darting out to take a worm, insect larva or small fish, or to drive away an intruder.

Fat fin mystery

In salmon, trout and other members of the salmon family, the second dorsal fin has been modified to what is called an adipose fin. This is a small flap made up of fatty tissue covered by skin and lacking fin rays or any other supporting skeleton. Some members of the large freshwater family of characins (Characidae) also have an adipose fin, as do the majority of catfish (order Siluriformes), the size of the fin varying from large to small according to the species. In the armored catfish, *Callichthys callichthys*, the adipose fin has a strong spine in front.

Conjecture surrounds the function of the adipose fin. Biologists know that the front dorsal fin of a salmon or trout prevents the fish from rolling and yawing as it moves forward. The pectoral fins, which prevent pitching and rolling, are used in turning and, with the pelvic fins, for braking. The tail fin helps to drive the fish through the water and also acts as a rudder. No one has yet discovered what the adipose fin does. Some ichthyologists believe it has no function, although this is scientifically unsatisfactory.

TRUMPETER

A gray-winged trumpeter near Manaus, Brazil. Trumpeters feed by pecking and scratching for fruit and insects on the forest floor.

TRUMPETERS ARE BELIEVED TO BE a link between the rails and the cranes. About the size of a domestic chicken, up to 28 inches (70 cm) long, they look rather like dumpy cranes with a hunchbacked stance. The legs and neck are fairly long, the head is small and rounded and the bill is like that of a chicken. The wings are rounded, and the tail is so short that it is hidden under the feathers of the rump. The feathers of the head and neck are short and soft, almost like hair, and those of the body are long and loose like those of an ostrich. The plumage is mainly black, with a purplish sheen on the neck.

There are three species of trumpeters, all native to Amazonia in tropical South America. The gray-winged trumpeter, *Psophia crepitans*, is black with gray on the back; it ranges from Venezuela and the Guianas to eastern Ecuador and northeastern Brazil. The pale-winged trumpeter *P. leucoptera*, with white feathers over its rump, lives in Peru and northwestern Brazil. The green-winged trumpeter *P. viridis*, which is black with some green on the wings, is found in northern Brazil.

Rooster minder

Trumpeters perch but are poor fliers, preferring to run swiftly to evade capture, and they will even swim rather than fly to cross a wide river. Because they are said to be good to eat, they are becoming rarer as human settlements advance into the Amazonian forest. Those that survive in the depths of the forests are not easy to study in their natural state, and their habits are not well known to science. Trumpeters are said to be easily tamed, however. Local people keep them as a means of protecting their poultry; the trumpeters join the chickens and establish themselves at the top of the pecking order.

Voices for contact

Trumpeters live in flocks of up to 200, roosting together in trees. The members of a flock keep in contact with each other by loud, reverberating calls, which start with short notes and end with drawn-out cries. Imitation of these calls is a sure lure for trumpeters. The birds get their name not from these contact calls but from a loud trumpeting that is used when threatening other trumpeters. It is a deep-toned, ventriloquistic sound, and trumpeters used as watchdogs call noisily at night at the first sound or sight of an intruder. Close-up, the sound seems to come from the depths of the bird's body. One 18th-century writer compared the sound of the trumpeting to the "lengthened doleful noise which the Dutch bakers make by blowing a glass trumpet, to inform their customers when the bread comes out of the oven." They also have a quiet call, the function of which is not known.

Forest gleaners

Trumpeters feed on fallen fruits and berries, insects, such as ants and flies, and other arthropods. They occasionally take small vertebrates. Trumpeters have the habit of following toucans, howler monkeys and coatis, in order to feed on the fruits that these animals drop from the trees. Pale-winged trumpeters have been seen taking fruits directly from small plants. All seeds that are ingested with the fruit are defecated intact, and by dropping the seeds on their journeys through the forest, the birds facilitate the spread of new vegetative growth.

TRUMPETER

CLASS	**Aves**
ORDER	**Gruiformes**
FAMILY	**Psophiidae**

GENUS AND SPECIES **Pale-winged trumpeter,** *Psophia leucoptera* **(detailed below); gray-winged trumpeter,** *P. crepitans*; **green-winged trumpeter,** *P. viridis*

ALTERNATIVE NAME
Trumpet bird

WEIGHT
2⅕–3⅓ lb. (1–1.5 kg)

LENGTH
18–21 in. (45–52 cm)

DISTINCTIVE FEATURES
Large size; mainly black plumage, with white or ocher patch on hind wing; long, hunched neck; long legs

DIET
Ripe fruit, insects and other arthropods, small vertebrates

BREEDING
Breeding season: eggs laid September–April; number of eggs: 3; incubation period: 23–29 days; fledging period: chicks nidifugous (can leave nest on day of hatching); breeding interval: 1 year

LIFE SPAN
Probably 10–15 years

HABITAT
Mature, dense, moist forest, from lowlands to 2,500 ft. (750 m) altitude

DISTRIBUTION
Amazonia in South America: northwest Brazil, eastern Peru, northeast Bolivia

STATUS
Widespread and not threatened, though becoming scarce in range

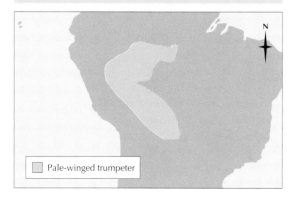

Pale-winged trumpeter

Exuberant courtship

In their courtship, large flocks of trumpeters gather in open spaces in the forest and perform energetic dances, leaping and calling and even turning somersaults. These appear to be unusual antics for such heavy-looking birds and are possibly an indication of their relationship with the cranes, which also conduct lively dances.

After mating, a pair of trumpeters selects a site for the nest, which may be a hole or fork of a tree or on the ground at the base of a tree. Three white eggs are laid. The incubation period lasts 23–29 days in *P. leucoptera*, 27 days in *P. viridis* and 28 days in *P. crepitans*. The chicks hatch with thick black down streaked with pink. Those that hatch from nests on the ground start to run about with their parents soon after hatching.

Trumpeters' tubes

Several birds have received the name of trumpeter from their loud, reverberating calls. These include the trumpeter manucode (*Phonygammus keraudreni*), an Australian bird of paradise, and the trumpeter swan (*Cygnus buccinator*). They owe their loud, trombonelike calls to an extremely long windpipe. That of the trumpeter and the trumpeter manucode is coiled under the skin of the breast. Other birds with long windpipes include the whooping crane (*Grus americana*), which possesses a 5-foot (1.4-m) windpipe that is coiled within the breastbone, and the plains chachalaca (*Ortalis vetula*).

The windpipe acts as a resonator, in much the same way as the pipe in an organ. As in the latter, the pitch of the sound that the windpipe can produce depends on its length and diameter.

The dark-winged trumpeter has iridescent plumage on its underparts. It inhabits the deep forests of the Amazon Basin, and is hunted locally for its flesh.

TRUNKFISH

A smooth trunkfish,
Lactophrys trigueter.
Bold skin patterns are a
feature of these fish,
many of which rely on
striking coloration to
deter predators from
tasting their toxic flesh.

THE TRUNKFISH ARE THE nearest approximation among fish to turtles. They are also known as boxfish and cofferfish because their bodies are enclosed within bony boxes, or carapaces, made up of six-sided plates that fit one another closely and leave only the tail unarmored. Inside, the backbone is short, with only 14 vertebrae between the skull and the beginning of the tail, all connected in a compact manner.

A typical trunkfish has a fairly conical head, the face sloping down at a steep angle to the small mouth, which is armed with strong crushing teeth. The eyes are large and there is only a small opening from the gill chamber. The length of a trunkfish seldom exceeds 1 foot (30 cm). The single dorsal fin and the anal fin are fairly large, as are the pectoral fins, but there are no pelvic fins. The fleshy, naked tail fin projects backward from the bony box and ends in a large, fanlike tail; apart from the other fins, it is the only part capable of movement. The box around the body has a flat undersurface. It may be three-, four- or five-sided in cross section, and one or more of its edges may possess strong spines. Trunkfish live on or near the bottom in warm waters, especially in tropical seas, worldwide.

Slow motion

As with tortoises on land, trunkfish are slow-moving animals, and for much the same reasons. A typical fish swims by strong side-to-side movements of the whole body, using its muscular tail. A trunkfish can move its tail only to a small extent. Its action is similar to that of a small boat being propelled by a single oar

sculling from the stern, though the process is somewhat more complex in the fish because the tail is flexible. The main swimming force is produced by side-to-side movements of the dorsal and anal fins, aided by the pectoral fins. A trunkfish is far from streamlined, particularly because the flat face creates resistance to progress. When swimming it moves its fins very rapidly, giving the impression of a great expenditure of energy with only a limited gain in forward movement.

Colorful warning

Rapid movement is not necessary for such a heavily armored a fish, which also can rely on its color for security and on its ability to poison predators. A common trunkfish found across the tropical Atlantic is the cowfish (*Acanthostracion quadricornis*), so named because it has two sharp, forward-pointing spines on the forehead, rather like the horns of a cow. It is pale green in color, marked with blue spots and lines, but it can change this to yellow with blue spots or brown with a network of pale blue markings, or even to pure white. The colors also differ between the sexes. The four-sided blue trunkfish, *Ostracion meleagris*, of the Indian Ocean and South Pacific is an example. The female and the young fish are purplish blue with numerous small white spots scattered thickly over the whole body. The male is very different, being purplish blue with a pale blue network except for the flat upper surface, which is a brownish purple peppered with small white dots with a brick-red border. Even the eyes differ: in the female and young fish they are blue, while in the male they have a red border.

Many biologists believe that these gaudy colors act as warning colors, advertising to possible predators that trunkfish do not depend entirely on their armor for defense.

Toxic defense

These fish secrete a poisonous substance known as ostracitoxin, which can kill not only other fish but also themselves. Glands in the skin, especially in the mouth region, produce the toxin, which appears to be exuded in greatest quantity when the trunkfish are under duress. When a trunkfish is placed in an aquarium, the other fish quickly begin to show signs of distress. They rise to the surface to gulp air, and usually soon die. The poison persists in the water even after the trunkfish has been removed. The only fish not affected are hardy species such as moray eels, the large groupers and some trunkfish.

Coral crunch

Trunkfish live among corals, biting off pieces of the coral to digest the soft-bodied polyps. As they do this, they expose worms and other small

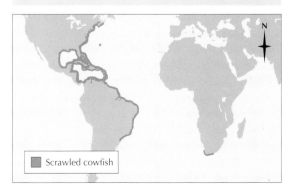

SCRAWLED COWFISH

CLASS	**Actinopterygii**
ORDER	**Tetraodontiformes**
FAMILY	**Ostraciidae**
GENUS AND SPECIES	***Acanthostracion quadricornis***

LENGTH
22 in. (55 cm)

DISTINCTIVE FEATURES
Compact shape; bony body in form of boxlike carapace; spine on each lower ridge of carapace; spine extending in front of each eye; yellowish coloration with blue spots; 2 to 4 blue bands on each cheek

DIET
Sessile (attached to substrate) invertebrates, including tunicates, gorgonians and anemones; sponges and sluggish crustaceans

BREEDING
Lays planktonic eggs

LIFE SPAN
Unknown

HABITAT
To depths of around 260 ft. (80 m), mainly in seagrass beds

DISTRIBUTION
Tropical and temperate waters of western Atlantic; southernmost South Africa (rarely)

STATUS
Not threatened

Scrawled cowfish

---

invertebrates sheltering in the coral heads. Some trunkfish use their spoutlike snouts to blow jets of water at the sandy bottom to uncover and dislodge worms, mollusks and small crustaceans, which they immediately snap up.

Dingleberries

The breeding habits of the cowfish of tropical American waters are probably typical of the whole family. It lays buoyant eggs, 0.8 millimeters in diameter, which hatch in 2–3 days. The larvae begin to develop a hard carapace in about a week. They become somewhat rounded in shape, and it is only as the young fish mature that the boxlike edges of the body become sharply defined. Young trunkfish have large eyes, small mouths and what appear to be puffed-out cheeks. In the United States, their rounded shape has earned them the name of dingleberries. During the early stages of life, they shelter under clumps of floating seaweed.

An unexpected delicacy

The heaviest mortality among trunkfish occurs in the early stages, when a great proportion of the eggs, larvae and young fish are eaten. Those that do reach maturity can rely on their protective boxes and, in some species, the poison they give out, to deter most predators. Being so slow, they lack the large muscles that make the flesh of other fish attractive to hunters. Yet trunkfish are eaten, even by humans, and in some places are regarded as a delicacy. They are cooked in their own boxes, and some peoples in the South Pacific traditionally roast them, rather like chestnuts. The liver is proportionately quite large and oily.

A spotted trunkfish, Lactophrys bicaudalis, among coral off Honduras in the Caribbean. Trunkfish and cowfish favor the warmth of tropical waters and are often to be seen around reefs.

TSETSE FLY

A tsetse fly pauses after feeding, its abdomen distended and weighed down with blood and its piercing mouthparts hinged forward in the inactive position.

THE NAME TSETSE IS probably of Bantu origin and is given to a genus of flies, *Glossina*, endemic to Africa south of the Sahara. They are robust and bristly-bodied, usually yellowish or brownish in color. About 20 species are known. A little larger than houseflies, *Musca domestica*, tsetse flies differ most obviously in the way that they fold their wings, scissorlike, over their backs when at rest. *Glossina* belongs to the same family as the housefly (Glossinidae) and is also closely related to the common European biting stable fly *Stomoxys*. The mouthparts of these flies are modified for piercing and sucking; so, too, are those of the tsetse flies.

Tsetse flies have become notorious as carriers of the disease known as nagana in cattle and horses and of sleeping sickness in humans. They feed by piercing the skin of a mammal, bird or reptile with their mouthparts and sucking the blood. Unlike mosquitoes, in which only the female sucks blood, both sexes of the tsetse fly feed in this way.

Each species of tsetse fly has its own habitat preference in sub-Saharan Africa. *G. palpalis*, the most important carrier of sleeping sickness in humans, favors dense forest bordering rivers and lakes. It lives largely on the blood of reptiles, such as crocodiles and monitor lizards, and also on that of the sitatunga (*Tragelaphus spekei*); it also bites humans. In open forest or savanna, *G. morsitans* and some related species depend on game animals for food. These are the carriers of the nagana cattle disease and of one of the types of sleeping sickness.

High parental care

Most insects lay large numbers of eggs to compensate for the heavy mortality suffered by their larvae, which must fend entirely for themselves without the power of flight to escape from their numerous predators, such as birds, fish and lizards. Tsetse flies produce their offspring in a similar way to mammals. The female develops only one egg at a time and it hatches inside her body. The larva remains in her uterus for 9 days and is fed on a secretion produced by glands that open by a nipple near the larva's mouth. These are, in effect, milk glands. The female maintains the supply of this fluid by constantly taking meals of blood. The larva breathes by means of a pair of black knobs that reach to the exterior through the opening of the female fly's oviduct.

When it is fully grown, the larva is extruded, or born. It falls to the ground and immediately pupates in the soil, where it remains for several weeks before emerging as an adult. A female tsetse fly may live for about 6 months. Though physically capable of reproducing every 10 days or so, she normally gives birth to no more than 12 larvae during her life.

Tsetse flies and disease

Sleeping sickness, or trypanosomiasis, is a debilitating disease that runs a slow course, from a few months up to several years, and ends in coma and death. It is caused by infection from protozoans (single-celled organisms) known as trypanosomes. The trypanosome undergoes part of its cycle of development in the blood of the vertebrate host and another part in the fly. There are two forms of sleeping sickness, both of which are caused by various subspecies of *Trypanosoma brucei*. The type known as the Gambian form is caused by *T. b. gambiense*, carried mainly by *Glossina palpalis* in West and Central Africa. The East African form, caused by *T. b. rhodesiense*, is conveyed mainly by *G. morsitans* and is prevalent in East and southern Africa. East African sleeping sickness first appeared in 1909 in Zimbabwe (then Rhodesia) and is the more severe of the two, often causing death within a few months.

White man's grave

Early European explorers referred to Africa as the "white man's grave," a grim sobriquet that owed as much to the lethal capabilities of the

TSETSE FLY

PHYLUM	**Arthropoda**
CLASS	**Insecta**
ORDER	**Diptera**
FAMILY	**Glossinidae**

LENGTH
⅖–⅘ in. (10–20 mm)

DISTINCTIVE FEATURES
Similar appearance to housefly, but a little larger and bristly; fully developed piercing mouthparts; single row of hairs on arista (bristly appendage) of each antenna

DIET
Blood of vertebrates

BREEDING
Holometabolous (undergoes full metamorphosis) and larviparous (produces hatched larva); number of eggs: 1; larva hatches immediately; female extrudes larva after 9 days; larval period: several weeks

LIFE SPAN
Up to 6 months

HABITAT
Varies according to species. Moist riverine forest (*Glossina palpalis*, *G. tachinoides*) to open savanna (*G. morsitans*, *G. pallidipes*, *G. swynnertoni*)

DISTRIBUTION
Sub-Saharan Africa

STATUS
Common

A newly emerged adult tsetse fly perches near the pupal casing in which it developed from a larva. Tsetse flies are unusual in that the female produces only one offspring at a time.

inexpensive to set up, can be used to lure them into open country. They settle on the model and can then be destroyed. They can also be persuaded to deposit their larvae on prepared sites, after which the pupae can be destroyed. Fly screens over doorways are also an effective deterrent.

Sleeping sickness can be treated if diagnosed in its first stage, in which the protozoans remain in the lymph and blood system; in the second stage they enter the brain, and treatment is then less effective. The drugs traditionally used are suramin sodium or pentamidine, depending on which trypanosome is being treated, although the former is not widely available and the latter is now too costly for most African clinics to stock. In Zimbabwe, wholesale slaughter of game animals has been effective in controlling the East African sleeping sickness.

The cattle disease nagana is far more difficult to control. The protozoans *T. b. brucei* and *T. b. congolense*, among others, are carried by several species of tsetse flies, which attack a great variety of animals, birds and reptiles. In these circumstances, extermination of wildlife is both impractical and ethically insupportable.

Return of an ancient killer

Medical advances up to the mid-1960s saw incidences of sleeping sickness decrease, but the disease has made a comeback, due in large part to the political instability of newly independent nations and to the unavailability of effective drugs. Today tsetse flies occupy 4½ million square miles (11.7 million sq km) of Africa, and this, according to the humanitarian aid agency Médecins Sans Frontières, places around 60 million people at risk of catching sleeping sickness. By the end of the 20th century an estimated 150,000 to 300,000 people were dying every year from the disease. Along with the malarial mosquito, the tsetse fly remains one of Africa's biggest killers.

tsetse fly as to the malarial mosquito. While Europeans made inroads from the west and south of the continent, much of the interior lay beyond their reach until the 19th century. Even then the tsetse fly proved impossible to control. Bullock trains and horses fell foul of nagana, as did the herds of cattle introduced for meat.

Effective control

Gambian sleeping sickness can be controlled by felling riverside vegetation or by catching and trapping the flies in paths and clearings. The disease retreats as a matter of course as the increase of human population leads to agriculture replacing waterside forest. Tsetse flies, which find their victims by sight, are attracted by dark colors and repelled by white clothing. Thus, models of large mammals, crudely made and

TUATARA

THE TWO SPECIES OF TUATARAS are the sole survivors of the beak-heads, a group of ancient reptiles that flourished from 225–120 million years ago, when dinosaurs roamed the earth. Around 60 million years ago they became extinct everywhere except New Zealand.

Outwardly, tuataras look like medium-sized lizards, but internally there are major differences. They grow to a length of about 2 feet (60 cm), the male being larger than the female. They have a large head, with teeth set along the edges of the jaws and a pair of enlarged upper front teeth. There is no external opening to the ear. The body and limbs are powerful and there are sharp claws on the partially webbed, five-toed feet. There is a crest of enlarged spines along the neck, back and tail, and the back and sides of the body are covered with small, granular scales and tubercles. The scales of the underparts are larger and more regularly arranged.

Like many lizards, tuataras are able to regenerate their tails, but they do not do so as effectively as their modern counterparts. The body color ranges from blackish brown to olive green; some specimens are gray. The body is peppered with small cream or yellow spots. These can be quite bright in juveniles and adults that have just shed their skin, but tend to fade with age. The scales of the crest are cream-colored, sometimes green.

Three-eyed lizards

Perhaps the tuataras' most interesting feature to zoologists is the presence of the third, or pineal, eye, which is a feature of many fossil vertebrates. This pineal eye is not peculiar to the beak-heads. It is shared with many other species of lizards, but it is better developed in the adult tuatara than in any other animal. Its presence in many embryos confused early ideas of evolution's relationship to embryology. The pineal eye is situated on the top of the brain, with a hole in the skull just above it, and has a lens and retina but no iris. It connects with a glandular body in the brain. The pineal eye probably was an important sense organ in some of the earlier reptiles but its function in the tuatara remains uncertain.

One breath an hour

Tuataras may dig their own burrow or share one with any of a number of petrel or shearwater species. Naturalists presume they live only on islands where the topsoil has been so manured

The tuatara burrows in soil naturally enriched by the guano of nesting seabirds, helping to sustain the numbers of soil-dwelling invertebrates on which the reptile feeds.

TUATARAS

CLASS **Reptilia**

ORDER **Rhynchocephalia**

FAMILY **Sphenodontidae**

GENUS AND SPECIES **Tuatara, *Sphenodon punctatus*; Gunther's tuatara, *S. guntheri***

ALTERNATIVE NAME
Brother's Island tuatara (*S. guntheri*)

LENGTH
Male: 24 in. (60 cm); female: 20 in. (50 cm)

DISTINCTIVE FEATURES
Lizardlike, crested reptiles lacking external ear opening; many internal features distinguish them from true lizards

DIET
Mostly invertebrates; lizards; eggs, chicks and adults of shearwaters and petrels; own young

BREEDING
Age at first breeding: more than 20 years; breeding season: January–March; number of eggs: up to 19, laid October–December; incubation period: 12–15 months

LIFE SPAN
Up to 70 years

HABITAT
Grassy islands, sometimes with small trees, often surrounded by sheer cliffs, and with large numbers of breeding burrows of shearwaters and petrels

DISTRIBUTION
Around 30 islets off northeast coast of North Island and in Cook Strait, New Zealand

STATUS
Rare; strictly protected. Population estimated at 50,000 to 60,000, much of it sited on Stephens Island, including only 300 specimens of *S. guntheri*.

Nighttime finds the tuatara leaving its burrow to hunt. Unlike many reptiles, its periods of activity are not restricted to the hours of daylight.

and worked over by these birds' many burrows that there is a layer of loose upper soil 18–24 inches (45–60 cm) deep.

Tuataras often bask in the sun during the morning or evening but spend most of the day in the burrow, coming out to hunt for food only at night. They are active at quite low temperatures, sometimes as low as 45° F (7° C), the lowest temperature recorded for any reptile activity. Their rate of metabolism is correspondingly low. Their normal rate of breathing even when active about is only one breath per seven seconds, but tuataras can go for at least an hour without taking a breath.

Although it is good-tempered when handled gently, a tuatara scratches and bites in self-defense. Its voice is a harsh croak similar in sound to that of a frog.

The tuataras' diet consists largely of crickets, beetles and other insects, and spiders, although snails and earthworms are also eaten. Contrary to the popular image of bird and reptile living amicably together, tuataras sometimes eat petrel eggs and chicks, even an occasional adult bird.

Unusually long gestation

Pairing occurs during January, but the sperm are stored in the female's body until the following October to December, when 5 to 19 white, oval, hard-shelled eggs are laid in a shallow depression in the ground that has been scooped out by the female and covered over with soil.

The eggs, which receive no parental attention, hatch some 12–15 months later, the longest incubation period known for a reptile. The young tuataras, which are brownish pink in color, break the shells of their eggs and dig their way to the surface. They are about 4½ inches (11 cm) long at this stage, but their growth rate is very slow. Tuataras do not breed until they are more than 20 years old, and they continue to grow until they are 50.

The tuatara has been kept in captivity on a number of occasions, one specimen in New Zealand having been kept in excess of 50 years. In the wild it lives to more than 50 years, and some claim that it may live to be more than 100.

Island sanctuaries

Tuataras were common on parts of mainland New Zealand until settlers came. Centuries ago the Maori arrived from Polynesia, bringing the Polynesian rat, *Rattus exulans*, although it was the appearance of Europeans in the early 19th century that spelled the end of mainland tuatara populations. These settlers brought not only their livestock, which grazed and trampled the native vegetation, but also the black rat (*R. rattus*) and Norway rat (*Rattus norvegicus*), which ate tuatara eggs and young. These predators are absent from the islands to which the tuatara is now confined, and these are the only locations where it can live and breed without disturbance.

A Tuatara Recovery Plan, set up in 1990, is managed by the New Zealand Department of Conservation and a Tuatara Recovery Group. Its working methods have included the removal of eggs from captive females taken from Cuvier, Little Barrier and Stanley Islands. The hatched offspring are reared in captivity, then released back into the wild when they have grown large enough to cope on their own. Some are released back to their source island, while others are used to start new populations on predator-free islands. Although much of the captive-breeding work, especially that involving the most vulnerable island populations, has been conducted in New Zealand, a number of tuataras have been sent overseas, in particular to Australia, where genetic research is conducted.

New species comes to light

Fossils show that tuataras were once found on both the North and South Islands of New Zealand, but now they are restricted to 30 islets off the northeast coast of North Island and in Cook Strait. For many years it was thought that these belonged to a single species named *Sphenodon punctatus*. A survey of tuataras carried out in the southern summer of 1989–90, however, showed that the tuataras living on North Brother's Island in Cook Strait were sufficiently distinct from the remainder that they are a separate species. This is now called the Gunther's, or North Brother's, tuatara, *S. guntheri*. Surveys also indicate that a distinct subspecies of *S. punctatus* may exist on Stanley Island.

A tuatara among the leaf litter on Stephens Island, Marlborough Sound, New Zealand. This island has about 30,000 tuataras, the largest population of the lizards in the world.

TUBESNOUT

THE TUBENOSE AND TUBESNOUT are closely related to the sticklebacks, and also to half a dozen other fish with tubelike snouts. They are placed in the family Aulorhynchidae. Both species are marine fish, the former living in the western Pacific and the latter living off the Pacific coast of North America from Alaska to California, down to depths of 100 feet (30 m), but mostly at depths of 30–40 feet (9–12 m). These fish are slender in the extreme and up to 7⅕ inches (18 cm) in length, the head making up about 1¼ inches (3 cm) of this. There is a single triangular dorsal fin just in front of the slender tail, which ends in a small, slightly forked tail fin. A row of 24 to 27 small spines extends from just behind the head to the leading edge of the dorsal fin. The anal fin is opposite the dorsal fin and is similar in shape and size. The very small pelvic fins lie almost under the throat, level with the pectoral fins. The body is colored olive green on the back, shading to silvery on the flanks and whitish on the belly.

Monster shoals

Tubesnouts and tubenoses occur in coastal waters, typically among eelgrass and kelp beds. They spend all their lives in shoals or schools that can be very large. The largest on record was recorded in November 1950 off Santa Rosa Island in southern California, where biologists of the Fish and Game Commission measured a tubesnout shoal ¼ mile (400 m) in diameter, entirely filling the water between 30–70 feet (9–21 m) depth. Any estimate of the number of fish in this shoal must be largely a matter of speculation, but it could have been 200 million to 600 million, depending on how closely the fish were arranged in the shoal. Another huge shoal, comparable in proportions to this one but composed of the related 3-spined stickleback, *Gasterosteus aculeatus*, was recorded in an English river.

Aulorhynchids feed on small prey, such as amphipods, opossum shrimps, the zoea larvae of crabs and fish larvae. These are sucked into the slender, tubular mouth.

Seaweed nest

Aulorhynchids are unusual in that they spawn year-round. This same is true of the tubesnouts off southern California, but farther north breeding is seasonal, with reports of eggs and young in the water during April–June in British Columbia.

The male tubesnout builds a nest among the young growths at the base of a giant kelp, *Macrocystis*, or other seaweed. He binds stems of weed together using sticky, silvery threads given out from his kidneys, in the manner of sticklebacks. He then entices the female to the nest to lay her orange, translucent eggs. These are about 1/12 inch (2 mm) across and are only slightly

The tubesnout of eastern Pacific waters. Its long, slender snout hints at the fact that it is related to seahorses.

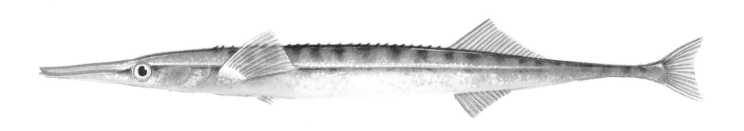

heavier than the surrounding water. They stick well to one another, although not to the kelp, but by clustering in rings around the stems they remain anchored to the plant, unless rough seas tear them away to hatch elsewhere.

Attentive father

For the next 2–3 weeks until the eggs hatch, the male guards the nest, driving away intruders. He also aerates the eggs by using his fins to fan currents of water through the nest. He prevents the larvae from leaving until they are well developed and are able to form small schools in the quiet waters close to the seabed, in the shelter of rocks or seaweeds. Egg-guarding by the male is not unique to this fish; it is also seen also among the lingcod, *Ophiodon elongatus*, and other greenlings, as well as gobies and sticklebacks.

The thickly growing fronds of kelp are a relatively safe nursery for the young tubesnouts, which prey on tiny, planktonic crustaceans and fish larvae. Tubesnouts mature and die in less than a year. The stomach contents of larger carnivorous fish, such as the lingcod, show that at times there are enough tubesnouts to serve as the main forage for predators. As a result, the swarms of tubesnouts often seen around the bays and estuaries of the Pacific coast are regarded as important to commercial fisheries.

Tubesnouts all

There are a number of fish with tubular mouths. The best known are the pipefish and seahorses (discussed elsewhere). Related to them are trumpetfish, family Aulostomidae, and the cornetfish, family Fistulariidae, named for the resemblance their mouths show to those musical instruments. The snipefish, family Macrorhamphosidae, are squat fish, with the head taking up about one-third of the total length and with the tubular snout recalling the long, slender bill of the bird after which they are named. Shrimpfish, family Centriscidae, are similar to them but have a transparent body wall, like that of a shrimp. The ghost pipefish, family Solenostomidae, also have a squat body, a flowing tail fin and an even longer tubular snout. Finally, discovered in 1929 in Lake Indawgyi, Myanmar (Burma), and the only freshwater species among them, is the fish *Indostomus paradoxus* of the family Indostomidae, which has been described as a cross between a pipefish and a stickleback.

All the above fish were once placed in an order far removed from the tubesnout. However, scientists have subsequently found that the species are closely related, linked by many features and showing all intermediate stages from a stickleback to a seahorse. In effect, all these fish might justifiably be termed tubesnouts.

TUBENOSE AND TUBESNOUT

CLASS **Actinopterygii**

ORDER **Gasterosteiformes**

FAMILY **Aulorhynchidae**

GENUS AND SPECIES **Tubenose, *Aulichthys japonicus*; tubesnout, *Aulorhynchus flavidus* (detailed below)**

LENGTH
Up to 7⅛ in. (18 cm)

DISTINCTIVE FEATURES
Small, extremely slender fish; soft, triangular dorsal fin set well back on body; small, finely forked tail fin; small, broad anal fin mirroring dorsal fin; truncated pectoral fins; pale mottled brown body, varying from olive green to yellow brown above and creamy white below; bright, silvery patch between gill cover and pectorals extending to throat and bounded above by dark band extending forward to snout; bright red, phosphorescent snout in breeding male

DIET
Small crustaceans, fish larvae

BREEDING
Builds nest, usually in kelp; territorial male guards nest, eggs and fry

LIFE SPAN
Less than 1 year

HABITAT
Shallow coastal waters with rocky or sandy beds; shoals around kelp beds, eelgrass and jetties, usually to depths of 40 ft. (12 m)

DISTRIBUTION
Eastern Pacific, from Sitka, Alaska, south to Punta Banda, Baja California

STATUS
Not threatened

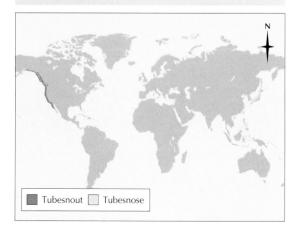

■ Tubesnout　□ Tubenose

TUBEWORM

OF THE MANY MARINE WORMS that live in tubes, *Chaetopterus*, the parchment-tube worm or tubeworm, is probably the most remarkable. Other tube-dwelling worms use sand grains or other hard materials to make their tubes. The tubeworm uses only the slime from its own glands to make a tube 15 inches (37 cm) to several feet long and ½ inch (12 mm) in diameter. The tube is made up of several layers of fine membranes. The worm, up to 10 inches (25 cm) long, has three parts to its body. The front part carries the mouth, two tentacles, 10 pairs of spear-shaped bristles and, behind these, a pair of winglike processes that can be brought together at their tips to form a ring that can be pressed against the wall of the tube. Behind this, on the upper surface, is a suckerlike cup lined with cilia (hairlike, mobile organs). A ciliated groove runs from the mouth to the cup. The middle part of the body is made up of five segments carrying three pairs of plate-shaped fans. The hind part of the worm, also the longest part, is more like the body of a ragworm (genus *Nereis*).

C. variopedatus is found all over the world, sometimes low on the shore, but more often just below the extreme low tidemark. Wherever it is located, it is likely to occur in high numbers. In some parts of the world it is specially numerous in the mud near the mouths of estuaries.

Food conveyor system

The mucous tube of the tubeworm is leathery and parchmentlike, and its color may be white, yellow, brown or green. It is U-shaped with only the ends protruding above the surface of the mud. These two ends are much narrower than the rest of the tube and are about 1 inch (2.5 cm) long. They are not conspicuous and can easily be missed, especially among kelp.

Other than when the worm is moving around in its tube, it is anchored to the wall of the tube by suckers on its undersurface. The three pairs of fans are used as paddles to drive water through the tube, bringing in oxygen and carrying away wastes. The current also brings food, which is trapped in a mucous bag. Mucus is given out from the winglike processes. Cilia in the groove carry the mucus back, stretching it like a bag until it reaches the ciliated cup. Because any water entering the tube must pass through this bag, any particles are trapped at its end. There, the cilia in the cup roll up the mucus and its contained particles to form a pellet. The action of the cilia in the groove is then reversed and the pellet is carried by them to the mouth,

where it is swallowed. Thereafter, another mucous bag is formed, more particles gather in it and another pellet is passed back to the mouth. The fans keep beating except when a pellet is being passed to the mouth. The interval between pellets averages 18 minutes.

Fine particulate feeder

The worm feeds only while it is in a tube. The moment it is taken out, it stops feeding, which makes it hard to determine the exact nature of its food, except by inference. However, its diet may include bacteria as well as fine particles of dead and decaying plant and animal matter. A related

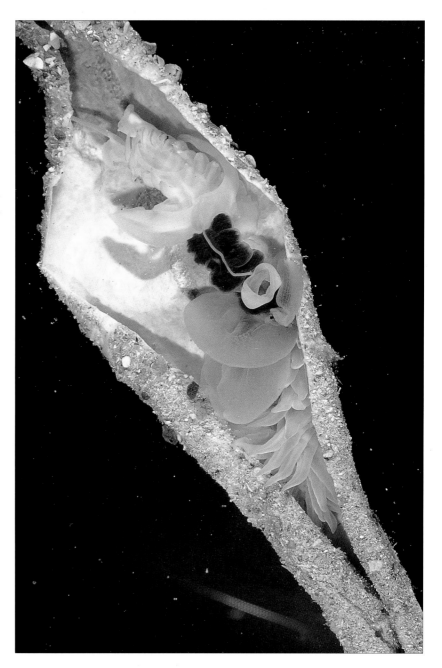

A cutaway photograph of a tube shows the species C. variopedatus, *head end uppermost. The dark body carries the ciliated groove, below which are the fans and tail section.*

Tubeworms projecting above the sand of a beach on Herm Island, off northern France. Each new tide brings more organic matter for the worms to filter out and digest.

TUBEWORM

PHYLUM	**Annelida**
CLASS	**Polychaeta**
FAMILY	**Chaetopteridae**
GENUS AND SPECIES	***Chaetopterus variopedatus***

ALTERNATIVE NAMES
Parchment-tube worm; innkeeper worm

LENGTH
6–10 in. (15–25 cm)

DISTINCTIVE FEATURES
Soft, fragile, yellowish or whitish body in 3 distinct parts; each body part made up of several segments; inhabits leathery, U-shaped tube, both ends protruding above surface; bioluminescent (emits light)s

DIET
Fine organic particles

BREEDING
Eggs and sperm shed into seawater, where fertilization occurs; group spawning common; free-swimming larva undergoes metamorphosis before settling on seabed

LIFE SPAN
Probably less than 5 years

HABITAT
Below or near low water in clean sand or mud of coasts, especially estuaries

DISTRIBUTION
Worldwide

STATUS
Not threatened

species, *Mesochaetopterus rickettsi*, for example, lives in a blind tube up to 8 feet (2.4 m) long that extends 4 feet (1.2 m) downward into the sand. One suggestion is that the blind end of the tube contains a well of stagnant water in which bacteria breed. In this species there is no mucous bag, but a string of mucus that is given off from the rear part of the body and continuously drawn along to the mouth. The mucus could be carrying bacteria with it. In other related species there is evidence that particles of dead plants and animals make up the food.

Orthodox life cycle

As among other marine tube-dwelling worms, breeding involves the shedding of eggs and sperm into the sea, where fertilization takes place. When one worm spawns, it liberates a chemical into the water that stimulates neighboring worms, so they all spawn together. From the fertilized egg a free-swimming larva develops. The first two parts of the young worm develop the most rapidly; the hind end is at first short with only a few tail segments instead of the 30 seen in the full-grown adult.

Renewing lost parts

The tubeworm can survive the loss of significant body parts, such as tentacles or the forepart of the head. It can regrow these lost parts in 10–14 days. It is also able to regrow the whole head end or most of the tail end. One tubeworm that broke up when being handled in the laboratory regrew both a head and a tail end from the middle portion of the three fans.

Shedding light on a puzzle

The tubeworm has given scientists a puzzle that has yet to be solved. Once it is out of the larval stage, the worm spends the remainder of its life in a tube buried in the mud. However, the moment it is disturbed, it lights up. If it is touched gently only on one part of the body, the light appears only on that part, but when the worm is handled roughly, the light spreads all over its body. The tubeworm has no eyes and so cannot use the light. The light may have evolved as a byproduct of the process by which oxygen is removed from the body, before aerobic respiration evolved in animals. Another possibility is that the light acts as an alarm system: any predator that attacks the worm is instantly lit up and risks being attacked by a larger predator.

TUBIFEX WORM

THE TUBIFEX WORM IS A SMALL, red aquatic worm and is used so often by aquarists as fish food that its scientific name has now become its common name. It is resembles a slender earthworm and measures up to 1½ inches (4 cm) long, although it can reach about 8 inches (20 cm) in length. Its body is cylindrical and divided into many ringlike segments. Four bundles of bristles protrude from each segment, except the first and last, with more than two bristles per bundle. These bristles are moved by muscles and are used to grip the mud when tubifex worms crawl and burrow. As in earthworms, there is a saddle, or clitellum, which is similar in appearance to a cigar band around the body and is involved in reproduction and the formation of egg cocoons.

Classification and species identification of tubifex worms is complex. One authority suggests at least 43 different scientific names have been used for this species. In addition, scientists have suggested many subspecies, forms and varieties. Tubifex worms occur worldwide; some live in fresh water and others in the sea, on shore or in shallow water, while a few live in estuaries or other waters of intermediate and fluctuating salinity. Of those living in the brackish water of estuaries, *Tubifex costatus* is essentially marine and *T. tubifex* is essentially a freshwater species. The common European species, *T. tubifex*, is found in both Europe and North America. There are also many related species, and those belonging to the genera *Ilyodrilus* and *Peloscolex* are sufficiently like the common species to be called tubifex worms.

Mud-grubbing bloodworms

In fresh water and estuaries, tubifex worms are characteristic of foul conditions, although they are not confined to them. They can flourish in water so lacking in oxygen that it supports no other life but sewage fungi and can be found in sewage filters, in sewage-fouled streams and in the mud of ponds. At low tide the mud of the Thames in London may be reddened by countless *Tubifex*, together with another red worm, *Limnodrilus*, of the same family, distinguished by the orangish yellow stripes on its rear end. There may be tens of thousands of *Tubifex* worms to 1 square yard (0.9 sq m) of mud. Another animal occurring in water nearly as low in oxygen as that which tubifex worms can stand is the blood red larva of the midge, *Chironomus*. The color of *Chironomus* larvae and *Tubifex*, both referred to sometimes as bloodworms and sometimes occurring together, is due to hemoglobin in their blood. This pigment helps them to make the best use of what oxygen there is available.

A tubifex worm lives with its head end downward in the mud, from which it digests bacteria and decayed organic matter. The worm is partly enclosed in a projecting, chimneylike tube of mud and mucus. However the tail end extends free in the water, writhing to and fro for much of the time to stir up the water and aid in the absorption of oxygen from it. As the oxygen content of the water falls, the worm tends to extend more and more of its length from its tube and may even come right out. Any disturbance of the water causes all the tubifex worms to pop back into the mud. If a number of worms are kept together without mud, they entwine themselves in writhing masses.

Reserve eggs for food

Tubifex is hermaphroditic, the male openings being on the underside of the eleventh segment from the front and the female openings in the groove between that segment and the twelfth. The clitellum is on both those two segments. Pairing tubifex worms exchange sperms cemented together in spermatophores, which are

Adult freshwater tubifex worms have between 112 and 130 body segments, each one except the first and last bearing four bundles of bristles.

When not embedded in their protective tubes in the mud, tubifex worms sometimes aggregate to form a squirming ball.

TUBIFEX

PHYLUM **Annelida**

CLASS **Oligochaeta**

ORDER **Tubificida**

FAMILY **Tubificidae**

GENUS AND SPECIES **Many species and subspecies, including *Tubifex tubifex* (detailed below)**

ALTERNATIVE NAME
Bloodworm

LENGTH
Up to 1½ in. (4 cm), usually less; may reach 8 in. (20 cm)

DISTINCTIVE FEATURES
Slender, blood-red worm; numerous bristled segments; head down in mud

DIET
Organic material and bacteria in mud

BREEDING
Hermaphrodite; pairs exchange spermatophores; numerous cocoons, each with 1 to 17 eggs; hatching period: several weeks

LIFE SPAN
Up to 2 years; normally about 1 year

HABITAT
Soft, organic-rich mud in freshwater rivers, streams, ponds and lakes

DISTRIBUTION
Ubiquitous in fresh water

STATUS
Common

elongated, glistening, white bodies up to ½ inch (2 mm) long. The spermatophores pass with the eggs into the cocoon that is secreted by the clitellum and there the cement is dissolved, releasing the sperms to fertilize the eggs. The whitish gray cocoons have a cylindrical neck at each end, are about 1/16 inch (1.5 mm) long and usually are oval. The eggs are 1/80–1/50 inch (0.3–0.5 mm) in diameter and there may be 1 to 17, usually 4 to 9, in each cocoon. In those cocoons with the most eggs, some may not hatch but serve as nourishment for the others. The eggs hatch after 8 to 56 days. The young worms are ¼ inch (6.3 mm) long and have 30 to 35 segments on hatching. There are 112 to 130 segments in the adult. Sexual maturity is reached in the fall, and cocoons are first visible in November, although the timing varies a great deal in northern Europe.

Relatives and predators

Most marine worms are polychaetes, meaning that on each segment there are many chaetae or bristles. Earthworms are oligochaetes, meaning they have few chaetae. The tubifex worm stands between them because, although it is an oligochaete, it has several bristles in each bundle. It also differs from an earthworm in having a gizzard, one pair of hearts to the earthworm's five, and in other anatomic details.

There are other relatives that burrow in the type of mud where tubifex worms are found, or that crawl on the seaweeds. Some use mud particles to make fixed or portable tubes, cementing them with slime produced by their bodies.

Its habit of exposing the hind end of its red body from the mud must at times make the tubifex worm very vulnerable to predators. The extreme conditions in which the worm flour-

ishes, however, generally do not support many other animals. One possible way in which marine worms might gain some protection was suggested by an Austrian scientist who collected some sand from the Mediterranean coast near Portofino for an aquarium. He later saw a red sea anemone sitting in the aquarium. The following day there were two anemones, each half the size of the original. On closer observation, the anemones proved to be collections of tubifexlike worms, although they were not positively identified as *Tubifex*, which were capable of creeping in a chain through the sand and reassembling at the surface elsewhere to form new "anemones." Small fish, normally fond of eating worms, clearly avoided swimming within their reach, just as they might avoid real anemones.

TUCO-TUCO

TUCO-TUCOS ARE HEARD more often than they are seen. Their common name is derived from their bell-like call note, which comes echoing up from underground, and though tuco-tuco is the usual rendition it also may be spelled tucu-tucu or tucu-tuco. Tuco-tucos are rodents, and in their outward appearance and habits they closely resemble the North American pocket gophers (discussed elsewhere) of the family Geomydidae, although tuco-tucos have no external cheek pouches.

Tuco-tucos are stocky animals. The head is large with a blunt snout and strong upper incisors, and the external ears are reduced to little more than small folds of skin around small openings on the sides of the head. The neck and limbs are short and muscular, with the forelimbs slightly shorter than the hind limbs. The hind feet are fringed with long hair, and the forefeet have strong claws that are used in burrowing and are longer than the toes. The fur varies in length and density among the various species, and the color ranges from dark brown to creamy buff. An adult tuco-tuco weighs 3½–24½ ounces (100–700 g) and can reach up to 14¾ inches (37 cm) in total length.

There are an estimated 48 species of tuco-tucos, but the number of species is constantly being debated. The species differ from one another only in minor points of appearance and habits, and many of them may be no more than local races. Tuco-tucos are found in tropical to subantarctic regions of South America from southern Peru to Tierra del Fuego, the largest numbers occurring in Argentina. In some areas they are becoming scarce, especially where the land has been fenced in for raising sheep or has been turned over to intensive agriculture.

Long shallow tunnels

Tuco-tucos are burrowers. They prefer dry sandy soil but are found in many different habitats, from coastal regions and plains to forests and high plateaus. One species, *Ctenomys lewisi*, tunnels in the banks of streams and it is thought that it may be partly aquatic. Tuco-tucos sometimes lives in large colonies, in which several adult females may occupy the same burrow. Biologists believe that in excavating their long tunnels these rodents use their head, chest and forefeet to loosen the soil and their hind feet to throw it behind them. The incisors are unusually

The tuco-tuco is a reticent animal. The high position of its eyes enable it to look out for danger without leaving the safety of its burrow.

broad and may be used in digging. Tuco-tucos groom themselves to remove loose sand from their fur by combing with the stiff bristles that grow near the bases of the hind claws.

The courses of tuco-tuco tunnels are marked by heaps of loose soil similar to molehills. The tuco-tuco is called the South American mole rat in some regions, and vicuñas (*Vicugna vicugna*) dust-bathe in sand thrown up by the animals' burrowing. The tunnels are fairly shallow, never more than a foot (30 cm) below the surface, and have chambers for food storage and nesting. The networks can be so extensive that a large area of land may become undermined, making horseback riding dangerous and hindering the movement of machinery. In some areas guinea pigs and lizards share the burrows with the tuco-tucos, and when the latter leave, mice and lizards often take them over.

Eyes in the tops of their heads

Tuco-tucos are most active early in the morning and again late in the afternoon, spending most of the rest of the day in their burrows. Before coming out to feed, they look out of the tunnel entrance to check if there is any danger. They can do this without showing themselves, as their eyes are almost level with the top of their heads. When feeding, tuco-tucos seldom wander more than a few feet from their burrows. For this reason they are seldom seen by the local inhabitants, although their calls are often heard. Unlike gophers, tuco-tucos have very good eyesight, and some species are believed to be able to distinguish a moving human as far as 50 yards (46 m) away.

Tuco-tucos feed entirely on plant foods such as grasses, tubers, roots and stems. One species, *C. opimus* in southern Peru, feeds almost entirely on spiny grass, *Festuca orthophylla*. Tuco-tucos collect food and hoard it in the chambers set aside in the burrows. They do not seem to need drinking water, suggesting that they probably obtain sufficient moisture from their food.

Variable mating season

The mating season for tuco-tucos varies widely. Gestation lasts 105–120 days, the precise period being species-dependent, and females produce up to seven young which are born in a grass-lined nest at the bottom of the tunnel. In *C. peruanus*, the young are well developed at birth and are able to leave the nest and feed on green vegetation almost immediately. They also can give the adult call. Tuco-tucos are able to breed before they are a year old but seldom live longer than 3 years in the wild. South American foxes, hog-nosed skunks, wild cats and hawks all prey on them.

TUCO-TUCO	
CLASS	**Mammalia**
ORDER	**Rodentia**
FAMILY	**Ctenomyidae**
GENUS	***Ctenomys***
SPECIES	**48, including *C. lewisi*, *C. opimus*, *C. peruanus***

ALTERNATIVE NAME
South American mole rat

WEIGHT
3½–24½ oz. (100–700 g)

LENGTH
Head and body: 6–10 in. (15–25 cm); tail: 2½–4¾ in. (6–12 cm)

DISTINCTIVE FEATURES
Thick gray or brown coat; loose skin across body; prominent orange-colored incisors

DIET
Plant matter, including grasses, roots, tubers and stems

BREEDING
Age at first breeding: 8 months; breeding season: varies according to species; number of young: 1 to 7; gestation period: 105–120 days, depending on species; breeding interval: probably 1 year (perhaps 2 litters per year in some species)

LIFE SPAN
Up to about 3 years

HABITAT
Most habitats, including subantarctic areas and tropical forests

DISTRIBUTION
Southern Peru to Tierra del Fuego

STATUS
Four species have declined greatly, including *C. magellanicus* which is endangered

☐ Tuco-tuco

TUNA

THESE LARGE FISH WERE known as tunny, a term deriving from the Latin *thunnus*, as early as the 15th century in England. It was during the 20th century that the word tuna, derived from Spanish, came into general use.

There are 14 species of tuna in five genera. The northern bluefin tuna, *Thunnus thynnus*, of the Atlantic can reach 15 feet (4.6 m) in length and weigh over 1,500 pounds (680 kg), but few exceed 8 feet (2.4 m). The yellowfin tuna (*T. albacares*) weighs up to 400 pounds (180 kg) and the albacore (*T. alalunga*) up to 80 pounds (36 kg).

The bluefin represents the typical tuna profile. It has a sleek, streamlined body with a large head and mouth and large eyes. The first dorsal fin is spiny, and close behind it is a smaller, soft-rayed second dorsal fin. The anal fin is similar in shape and size to the second dorsal, and behind these two, extending to the crescent-like forked tail, are finlets, nine on the upper tail root and eight on the lower. The pectoral fins are medium sized, as are the pelvic fins, which are level with the pectorals. The back is dark blue, the flanks white with silvery spots and the belly white. The fins are dark blue to black except for the reddish brown second dorsal fin and the yellowish anal fin and finlets. There are three distinctive keels on each side at the base of the tail fin. The bluefin is found across the North Atlantic as far north as Iceland.

Segregated by size

Tuna are oceanic fish that sometimes come inshore but apparently do not enter rivers. They move about in shoals in which individual fish are all approximately the same size. The larger the tuna, the smaller the shoal, and the largest individuals are more or less solitary. They swim near the surface in summer but are found at 100–600 feet (30–180 m) in winter.

Tuna are strongly migratory, their movements being linked with those of the fish on which they feed and on the temperature. They are intolerant of water temperatures below 50–54° F (10–12° C), so although they move into northern waters in summer, they migrate back to warmer seas in the fall. A cold summer limits the extent of the northward migrations. The fish also evidently cross the Atlantic. Tuna tagged off Martha's Vineyard, Massachusetts, in 1954 were caught in the Bay of Biscay, off the coast of Spain, five years later; individuals from North American waters occasionally turn up off the coasts of Norway. Two tagged off Florida in September and October, 1951, were caught off Bergen, Norway, 120 days later, having traveled 4,500 miles (7,200 km).

In common with mackerel, to which they are related, tuna swim with the mouth slightly agape so that their forward movement forces water across the gills. Their oxygen needs are high because of their great muscular activity, which depends on a correspondingly abundant supply of relatively warm blood. Tuna are unique among fish in their ability to maintain their blood at a temperature above that of the surrounding seawater. Owing to their high oxygen requirements, tuna swim more or less continuously, normally at a low cruising speed, though some scientists have reported brief bursts of up to 50 miles per hour (80 km/h).

Feeding frenzy

When they are very young, tuna feed on crustaceans, especially euphausians, but as they mature they eat mainly shoaling fish such as

The skipjack tuna, Katsuwonus pelamis, is found worldwide. It grows to about 3 feet (90 cm) and can weigh more than 45 pounds (20 kg).

herring, mackerel, sprats, whiting, flying fish and sand eels. They also eat some squid and cuttle-fish. When a tuna shoal meets a shoal of prey fish, it is seized with what may be described as a feeding frenzy. It charges through, twisting and turning, often breaking the surface and sometimes leaping clear of the water. The commotion usually attracts flocks of seabirds to feed on the smaller fish that are driven to the surface.

Putting on weight

In the Atlantic the bluefin tuna spawns in the Gulf of Cadiz, to the southwest of Spain, and in the Mediterranean in June and July. It spawns off Florida and the Bahamas in May and June.

A shoal of skipjack tuna in Pacific surface waters. Small tuna prey on crustaceans, such as shoaling shrimp, graduating to small fish as they grow.

TUNA

CLASS **Osteichthyes**

ORDER **Perciformes**

FAMILY **Scombridae**

GENUS AND SPECIES **14 species including northern bluefin tuna, *Thunnus thynnus* (detailed below); albacore, *T. alalunga*; yellowfin tuna, *T. albacares***

ALTERNATIVE NAME
Tunny

WEIGHT
1,500 lb. (680 kg)

LENGTH
15 ft. (4.6 m)

DISTINCTIVE FEATURES
Dark blue back, shading to green on upper flanks; silvery-white lower flanks and belly; first dorsal fin yellow or bluish, second dorsal fin reddish brown; dusky yellow anal fin and finlets, edged with black

DIET
Crustaceans; smaller fish, including herring, sprat, pilchard, whiting, sand eel

BREEDING
Breeding season: June–August (Atlantic), April–June (western Pacific); young spend first year in breeding area; mature at 5–6 years

LIFE SPAN
15 years

HABITAT
Oceanic but ventures close to shore; seldom found in waters deeper than 330 ft. (100 m)

DISTRIBUTION
Atlantic Ocean; Mediterranean and Black Seas

STATUS
Not known, owing to migratory habits

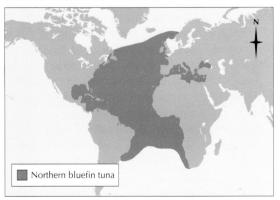

Northern bluefin tuna

Each year an adult female bluefin tuna may produce around 10 million eggs. These are small and float near the surface. They hatch in about 2 days, the new larvae being less than ¼ inch (6 mm) long. The fry grow quickly, weighing 1 pound (450 g) by 3 months and 10 pounds (4.5 kg) at a year old. At 13 years of age they reach a length of 8 feet (2.4 m) and weigh 440 pounds (200 kg).

Ancient fisheries

The many references to the tuna in classical literature reveal it to have been as important to the Mediterranean people as the herring was to the people of northwestern Europe. The fisheries have continued through the centuries. Catching methods have included harpoons, baited hooks and nets. Today, very long nets are used to intercept migrating shoals and guide them into a final compartment. When this is filled with jostling fish, the net floor is raised, the boats close in and the massed fish are clubbed, speared and dragged into the boats. This method became mired in controversy during the late 20th century because an unacceptably high number of dolphins, which habitually follow tuna shoals, became fatally ensnared in tuna nets. The globally important tuna fishing industry was pressed to adopt more dolphin-friendly fishing techniques, such as rod-and-line fishing.

Sport fishing for bluefin tuna also became very popular during the 20th century. A mature fish, played with a rod, is said to give the fisher the ultimate contest, perhaps towing a boat for hours over a distance of several miles before becoming exhausted. The tuna's chief natural predator is the orca or killer whale, *Orcinus orca*.

Wide-ranging tuna

There has long been doubt as to whether the tuna of the American Atlantic is the same species as that on the European side. The fish differ slightly in details of anatomy and in breeding habits. Nevertheless, the tendency now is to treat them as separate populations of a single species.

Other related species have similar wide distributions. A near relative, the Atlantic albacore, up to 4 feet (1.2 m) in length and with long, scythelike pectoral fins, has its counterpart in the Pacific albacore, which ranges from the Pacific coast of North America to Japan and Hawaii. In the yellow-finned albacores or yellowfin tuna, up to 9 feet (2.7 m) long and 400 pounds (180 kg) in weight, the second dorsal and anal fins are also long and scythelike. One species ranges across the tropical and subtropical Atlantic, the other across the Pacific and into the Indian Ocean. It is these fish that make up the bulk of the tuna caught for human consumption.

These bluefins are in a South Australian fish farm. Along with the yellowfin tuna, they have become an immensely valuable commercial species.

TUNDRA

The brief spring and summer months are a boon for herbaceous plants, which provide rich grazing for the herds of caribou that have come north after the winter cold.

IN THE ARCTIC, THE VAST boreal (northern) forest called the taiga fades gradually to the north as, one by one, tree species retreat from the harsh conditions. Beyond the forest lies the tundra, a treeless area stretching to the remote northern coasts of Alaska, Canada, Scandinavia and Russia. In Greenland and on Spitsbergen the tundra meets areas of permanent ice cap. In the Southern Hemisphere it is found on remote subantarctic islands and clings onto those coastal areas of the Ross Peninsula of Antarctica that are free of permanent ice. There are equivalent biomes on the upper slopes of mountains, above the tree line and below the region of permanent snow, and this is often called alpine tundra, as opposed to the Arctic tundra.

Arctic tundra regions are at such high latitudes, generally above 70° N, that the sun's energy strikes the land at an oblique angle, and the available energy is spread across a much wider area than in the Tropics. At these latitudes, there are extreme changes in day length from one season to the next. In summer, the sun dips briefly below the horizon, resulting in a short twilight that serves as night. In midsummer, the sun never sets at all. By contrast, the winter nights are long, the sun rising only briefly above the horizon. With these seasonal extremes comes an extreme climate, and the few plants and animals that survive in the tundra must have significant specializations.

Stunted growth

The winters are not only dark but also long. The average temperature is below 32° F (0° C) for 6–10 months of the year. Conditions are ideal for the creation of permafrost, a layer of permanently frozen ground that may extend to a depth of 2,000 feet (650 m) or more. It is the permafrost, among other things, that is thought to restrict the growth of trees. Even when the topmost layers of the soil are thawed in spring, which comes in June, the permafrost below prevents the free drainage of water and stops the penetration of deep tree roots. Biological activity can only take place in the surface layers.

The bitter winds that constantly lash the Arctic have a damaging drying effect and carry with them abrasive ice crystals. No plants can stand in the teeth of these gales, and those that live in the tundra grow huddled in low-lying mat and cushion formations, no more than 4 inches (10 cm) high, as far as the eye can see. If stream valleys cut deeply through the tundra, hardy little dwarf willows and birches can grow to some 10 feet (3 m); but as a rule, the tundra is apparently featureless, flattened by the ice sheets of the last ice age, and resembles a drab, green-brown grassland. Dwarf willows grow in the open, too, but there they grow only to 4 inches (10 cm) tall, making themselves obvious among the mosses and lichen only by their fluffy, catkinlike flowers.

Sturdy vegetation

There are more than 1,000 tundra-specialized plant species, but many of them are also found in regions far to the south of the tundra, and in isolated patches of alpine tundra. They include mosses, such as *Sphagnum*, and lichens, including the reindeer lichen, *Cladina rangiferina*, which is usually (and incorrectly) called reindeer moss. Most common are sedges, Cyperaceae, and cotton grass, which is also a kind of sedge.

Tundra plants are usually slow-growing perennials that gradually build up food reserves over years. Such plants must not only tolerate the difficult conditions above ground, but also the waterlogged, peaty soil conditions created by the permafrost. Despite the freezing and flooding, the soil seems to be a more favorable environment for plant growth than the air above, because the biomass of plants beneath the surface (roots and hidden shoot tips) exceeds that above the surface many times over. When even the top layers of the soil are frozen in winter, many tundra plants retreat below the surface, allowing their exposed parts to die off. Underground they are insulated somewhat from extremely low temperatures, and there they can protect their delicate growing tips.

In the brief 6–10-week growing season, the diverse and colorful flowering plants of the tundra quickly bloom, making the most of the long days of sunshine and the abundant meltwater. They include many heathers, Ericaceae,

TUNDRA

CLIMATE
Extreme, with brief growing season of 6–10 weeks (June to August or early September); average annual rainfall: 5–14 in. (130–350 mm); average maximum summer temperatures: 50° F (10° C); average winter temperatures: -5 to -22° F (-20 to -30° C), but can drop below -70° F (-57° C); long, dark winter creates conditions for permafrost layer in ground

VEGETATION
Mainly in mat or cushion form, up to 4 in. (10 cm) high. Plants include mosses, liverworts, lichens; sedges and grasses; heathers (Ericaceae), including deciduous rhododendron, blueberry, and cranberry; woody shrubs, including dwarf birch, dwarf willow; perennial herbs, including saxifrage

DISTRIBUTION
Northern Hemisphere: northern Alaska and Canada, coastal areas of Greenland, much of Iceland and Spitsbergen, northern fringes of Scandinavia, northern Siberia. Southern Hemisphere: coastal areas of Ross Peninsula of Antarctica; subantarctic islands, (e.g. South Georgia)

STATUS
Exploitation of oil and mineral reserves is beginning to threaten tundra ecosystems; vehicles and machinery scar land, and pipelines interfere with animal migration routes; climate change may allow taiga to encroach northward on tundra

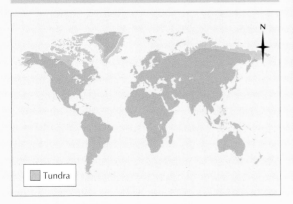

Tundra

Collared lemmings, which are found on the tundra, are the only rodents to turn white in winter, and they have a thick coat all year round. In summer they bulk up on fruit, flowers, sedges and grasses, switching to buds and bark in winter.

such as Arctic species of rhododendron, blueberry and cranberry. Their short-lived flowers are large, competing for the attentions of pollinating insects.

Hardy animals

Most animals deal with tundra conditions in one of two ways: they avoid the winters by migrating, or they endure them by partially escaping underground. Those that migrate may travel hundreds of miles south in winter, following what meager grazing they can find. Caribou, *Rangifer tarandus*, the domestic relatives of which are known in Eurasia as reindeer, can subsist even in winter by grazing on reindeer lichen. In winter, they move south to the edge of the taiga, but in summer they range over the northern tundra, even as far as the distant Arctic coast. The movement of caribou is at least partly controlled by mosquitoes and blackflies. Insect larvae develop very quickly in the brief summer and hatch from the bogs and lakes in unimaginable numbers, so dense that they appear like plumes of smoke. Once the mosquitoes have emerged in midsummer, caribou must confine themselves to poorer grazing on higher ground, away from river valleys, or they would be bitten to death by the insects.

For most animals living permanently in the tundra, the winters are too long and cold for hibernation to be viable. Also, the summer is too brief to enable them to put on sufficient body fat. These animals must stay active and able to feed themselves even in the depths of winter. The ptarmigan, genus *Lagopus*, does so by feeding on seeds and berries buried beneath the snow. Like other birds of the tundra, it grows white plumage at the end of summer for camouflage against the snow. Arctic foxes, Arctic rabbits and snowshoe hares grow white winter coats. The ptarmigan is prey for the snowy owl, *Nyctea scandiaca*, another year-round resident. The main prey of the snowy owl is the lemming, genus *Synaptomys*, a burrowing rodent that survives by tunneling in the snow and sheltering there, eating what vegetation it can find.

Herds of up to 200,000 caribou in northern Canada travel as far as 600 miles (1,000 km) between their summer tundra feeding grounds and their forested wintering grounds.

Migratory destination

The summer in the tundra, although brief, is a period of intense activity and productivity. The resources attract animals that are not hardy enough to survive the Arctic winter, but are mobile enough to reach the tundra from their wintering grounds hundreds of miles to the south. This includes strong-flying migratory birds, chiefly swans, geese and duck and waders such as plover. The tundra swan, *Cygnus columbianus*, is one of the many species that join a rich assemblage of birds traveling to the tundra to feed and breed. Many birds rely on the tundra for habitat in which to court, mate, nest and rear their young. Some birds graze the tundra vegetation, but the extensive wetlands created by the permafrost provide breeding ground for insects. These are preferred by the wading birds as a richer food source. The birds achieve a cycle of breeding within the brief Arctic summer, after which they return to their wintering grounds.

Fragile ecosystem

The remoteness and inhospitableness of the tundra has largely saved it in the past from the damaging activity of humans. However, pollution in the atmosphere can travel far, and the tundra seems to suffer badly from the global problems of acid rain and ozone depletion. More recently, humans have attempted to develop the tundra for oil and mineral exploration. This has revealed how fragile the tundra is to disturbance. The peaty soil overlying permafrost is damaged by heavy machinery. It bears the scars of tire tracks for decades after a vehicle has passed, because of damage to the frozen soil and because the plants cannot regenerate quickly. Food chains in the tundra are short, and food webs are simple, involving few species. If one species is adversely affected by pollution or other human activities, the balance of the ecosystem could be dramatically changed.

Oil prospectors have begun to invade the wilderness of the Arctic National Wildlife Refuge in northern Alaska, home to herds of caribou and even polar bears in winter. Meanwhile, northern Siberia has for some years been opened up for development. Human activity there threatens to melt the permafrost and cause the land of low-lying areas, such as the Yamal Peninsula in northern Russia, to subside into the sea. Global warming may contribute to this problem. If the situation continues to deteriorate, increasing temperatures may see the taiga encroaching farther north, ultimately reaching the Arctic Ocean and forcing the tundra off the map.

After wintering in Central and South America, the pectoral sandpiper, Calidris melanotos, *heads for the tundra to breed. The northern summer offers rich insect pickings.*

TURACO

Turacos are confined to Africa south of the Sahara. Some are widely distributed, such as the giant turaco of West Africa, the Democratic Republic of Congo and Malawi, but a few are very restricted, including the Prince Ruspoli's turaco, *Tauraco ruspolii*, which lives in a small area of Ethiopia.

Reversible toes for agility

Turacos are found only near trees, from the thick evergreen forests, where the giant turaco and Hartlaub's turaco, *T. hartlaubi*, live, to the dry savannas of the east coast of Africa. The Ruwenzori turaco, *Ruwenzorornis johnstoni*, lives up to 12,000 feet (3,600 m) above sea level. The forest-dwelling species have the most green in their plumage, whereas those living in thorn scrub may have none. In this way the forest turacos are well camouflaged, their red wing patches showing only in flight. If they are disturbed they freeze, becoming quite inconspicuous, and then quietly run through the branches, making their way up into the safety of the canopy.

Branches and foliage present no obstacles to turacos because of the distinctive form of their feet. When birds' feet feature two toes facing forward and two facing backward, as in owls and woodpeckers, they are known as zygodactylous. The feet of turacos are semizygodactylous; they have a special joint on the outer toe. At rest this toe sticks out sideways, but it can be moved either forward or backward. This special joint makes it easy for turacos to walk to the tips of the smaller branches, among the leaves and twigs where they hunt for their food.

Turacos are often gregarious and gather in groups where food is abundant. The yellow-billed turaco, *T. macrorhynchus*, roosts huddled in small groups, the presence of which is often given away by their loud, raucous calls, often occurring in chorus. Turacos are inquisitive birds and readily approach humans or snakes. Their calls are harsh and the go-away birds are named after their cries of *go-waa*.

Eating poison

The main food of turacos is fruit, and they show definite preferences for certain kinds. Turacos eat very wastefully and drop more than they consume, although the fallen fruits are presumably

In common with other turacos, the purple-crested turaco feeds primarily on fruit and is found in South Africa.

THE TURACOS, OR TOURACOS, some of which are known as louries, go-away birds or plantain-eaters, are lively, fruit-eating birds. Their nearest relatives seem to be nonparasitic cuckoos, such as the couas of Madagascar. There are 23 turaco species, most of them 16–21 inches (40–53 cm) long, the size of a wood pigeon, but the giant or great blue turaco, *Corythaeola cristata*, is 2½ feet (75 cm) long, about the size of a pheasant. The wings are short and rounded and the tail is rather long. The bill is strong and curved. The plumage usually is gray or brown, as in the go-away birds, but some species have considerable amounts of green with red patches on the wings and red, white or other colors on the head. Ross's turaco, *Musophaga rossae*, and the giant turaco are predominantly blue, while the purple-crested turaco, *M. porphyreolopha*, has violet wing coverts. All turaco species have crests except for the violet plantain-eater, *M. violacea*, which has short, hairlike feathers on the head, and most species have skin around the eyes or elsewhere on the face.

eaten by other animals. The birds also eat shoots and leaves, while some species also take small invertebrates. The black-billed or Congo turaco, *T. schuttii*, feeds on small snails, and the yellow-billed turaco feeds on invertebrates flushed out by driver ants on the forest floor.

ROSS'S TURACO

CLASS	**Aves**
ORDER	**Musophagiformes**
FAMILY	**Musophagidae**
GENUS AND SPECIES	***Musophaga rossae***

WEIGHT
About 17½ oz. (500g)

LENGTH
20–21 in. (51–53 cm)

DISTINCTIVE FEATURES
Dark body and tail; violet blue plumage; bright crimson crest; yellow bill and frontal shield; brilliant red primary flight feathers

DIET
Fruits, flowers, young shoots; termites; snails

BREEDING
Age at first breeding: 1 year; breeding season: eggs laid year-round; number of young: 2 or 3; incubation period: 21–24 days; fledging period: 10–12 days; breeding interval: 1 year

LIFE SPAN
Not known

HABITAT
Woodland; forest edge; riverine gallery forest; avoids deep forest

DISTRIBUTION
Angola and Zambia north to southern Sudan and northern Cameroon

STATUS
Locally common

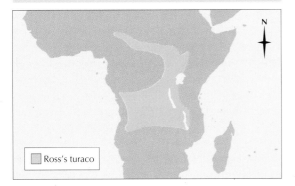

Ross's turaco

Precocious young

The nests of turacos are similar to those of pigeons: flat platforms of loosely woven twigs built in dense foliage, usually quite low in the tree. Most of their nesting habits have not been well documented but the ornithologist V.G.L. Van Someren has studied Hartlaub's turaco in detail. It has two breeding seasons in each year: April–July and September–January. However, it appears that each individual nests only once a year. Turacos living near the equator nest year-round, but elsewhere they generally do so just after the rainy season.

The floor of the nest is so thin that the two white eggs can be seen through it. The incubating parent holds its head in such a way as to break its silhouette and so appear less conspicuous. Ornithologists estimate the incubation period to last for 21–24 days and the fledgling period for 10–12 days. The latter is difficult to record because the chicks, which hatch with a covering of black down, leave the nest before they can fly, and clamber about using wing claws similar to those on young hoatzins or mousebirds. The chicks are fed on regurgitated fruit.

Unique pigments

The red and green colors in the plumage of turacos are unusual. In most birds, green is produced either by the structure of the feather or by a mixture of two pigments: brown melanin and a yellow carotenoid. In turacos, the green is created by a single green pigment named turacoverdin. The red pigment of the wings and head is called turacin.

For a long time, common belief held that the turacos' red pigment ran easily and that their red plumage faded in the rain. In fact, in museum specimens the red plumage darkens rather than fades, although such darkening with exposure is unusual in nature.

A native of coastal South Africa, the Knysna touraco, T. corythaix, is remarkable in that it eats the poisonous red fruits of a shrub known as bushman poison.

TURBOT

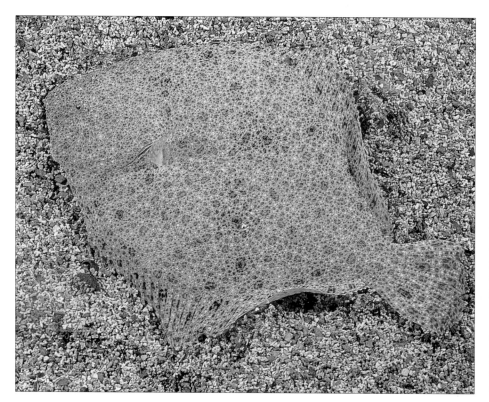

The turbot, Scophthalmus maximus, *varies its color to match the seabed. Its spots and speckles help to further camouflage it.*

THE TURBOT HAS THE reputation among many people of having the finest flavor of any sea fish. Ever since a fisherman presented a very large turbot to the Emperor Nero, the fish has had the reputation of growing to a considerable size. It was said there was no dish large enough to take it and that Nero summoned his senators to gaze on this piscine marvel.

The turbot grows up to 40 inches (100 cm) long, and its maximum weight is 55 pounds (25 kg), with the female being larger than the male. Its body is broad and diamond-shaped, the length being only 1½ times its width. Both eyes are on the left side, so the fish rests on the seabed on its right side. The large mouth is situated to the left of the eyes, and the teeth and jaws are equal on both sides. The dorsal fin starts at the snout and, like the anal fin, does not join with the tail fin. The pelvic fins are broad at the base, the one on the eyed side being slightly longer than the one on the blind side. The turbot is scaleless but its upper side is covered with small, scattered bony knobs or tubercles, which are much smaller and closer together on the head. The color varies according to that of the seabed on which it is lying, ranging from a dull, speckled and spotted sandy-brown when on mud or muddy sand to a pale yellow on sand. The underside is white but may bear patches of color. A few turbot are colored on both sides.

The brill, *Scophthalmus rhombus*, is closely related to the turbot and has a similar form and similar habits, but it is comparatively unimportant as a food fish. The Black Sea turbot, *S. maximus maestica*, is a subspecies of the Mediterranean turbot, and has much larger tubercles on both the lower and upper surfaces of its body.

The turbot ranges from the Mediterranean to the North Sea as far as Bergen on the Norwegian coast, sometimes wandering north of the Arctic Circle as far as the Lofoten Islands. It is found all around British coasts, but is rarely caught as far north as the Orkneys and Shetlands off the northern Scottish coast.

Rippling swimmer

The turbot lives in shallow water, rarely being caught in depths of more than 262 feet (80 m). Locally it is sometimes found so close to the shore that it can be fished off the beach at low water, but it has not been known to enter estuaries. It normally lies on shell gravel or gravel bottoms but also may be found on sand or mud. Like all flatfish, the turbot swims by rippling undulations of the whole body, but each spurt does not carry it very far.

A wide-ranging diet

Turbot feed mainly on other fish such as sand eels, sprats, pilchards and members of the cod family. Soles, dabs, dragonets, sea bream and boarfish, a relative of the John Dory, are sometimes taken. Turbot have also been noted to feed on invertebrates such as bivalve mollusks and worms, but only rarely. The larvae of the turbot feed on animal plankton, including the larvae of barnacles and mollusks.

Prolific spawners

In the North Sea and the Irish Sea, spawning takes place from April to August; in the western Channel and off the Welsh coast it occurs from May to September. The turbot is one of the most prolific of seafish, the female laying 5 million to 10 million eggs, each ⅟₂₅ inch (1 mm) in diameter, over gravelly ground in shallow water. Turbot eggs and larvae are pelagic (oceanic), floating near the surface of the sea. The egg has a pale yellow oil droplet in it that gives buoyancy. The eggs hatch after 7 to 9 days. At first the fry are

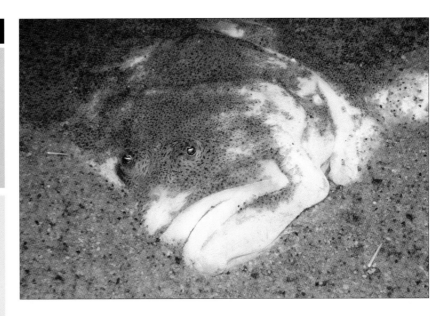

TURBOT

CLASS	**Actinopterygii**
ORDER	**Pleuronectiformes**
FAMILY	**Scophthalmidae**

GENUS AND SPECIES **5 genera and 18 species, including *Scophthalmus maximus* (detailed below)**

WEIGHT
Up to 55 lb. (25 kg)

LENGTH
40 in. (1 m)

DISTINCTIVE FEATURES
Flat fish; both eyes on left side of large head; wide body; color matches seabed

DIET
Mainly bottom-living fish (sand eels, gobies); some crustaceans and bivalves

BREEDING
Age at first breeding: not known; breeding season: April–August; number of eggs: up to 10 million; hatching period: 7–9 days; breeding interval: not known

LIFE SPAN
About 15 years

HABITAT
Shallow inshore waters on shell gravel, sandy, rocky or mixed bottoms

DISTRIBUTION
Northeast Atlantic, Mediterranean Sea, European coasts to Arctic Circle, Baltic Sea

STATUS
Not threatened

Scophthalmus maximus

the head to a position next to the left eye. The left side of the body becomes pigmented and the right side becomes white. The change is slow and does not begin until the young turbot is ½ inch (1.3 cm) long and is sometimes not completed until it is 1 inch (2.5 cm) long, and 4 to 6 months old. During this time the baby turbots are widely spread. They do not assume a bottom-living life until after their metamorphosis is complete.

Turbots grow fairly rapidly, the females faster than the males. At 3 years old the males are about 12½ inches (31.3 cm) long and the females 14½ inches (36.3 cm), and at 5 years old the males are 16½ inches (41.3 cm) long and the females 17½ inches (43.8 cm). The maximum age recorded is about 15 years, when the average length of males is 21 inches (52.5 cm) and the females 27 inches (67.5 cm).

Hazardous beginnings
As the eggs and larvae float near the surface of the sea, they are subject to many hazards. They are at the mercy of wind and current and may be carried into waters where they cannot survive, or they may be snapped up by a wide variety of fish. Changes of temperature may kill them or they may be cast up on the shore to die. Out of each brood of 10 million eggs, very few turbot reach adulthood. Once fully grown, the fish's ability to assume the color of the seabed where it is resting acts as an effective camouflage concealing it from predators.

Its fine flavor makes the turbot an important food fish in both the Mediterranean and northern European waters. The chief fishery is in the central area of the North Sea, where 88 percent of the northern European catch is obtained. Fishing now is usually by trawl, although some are taken by longlines or even caught in seine nets off the beach in some localities.

Newly hatched turbot fry have a regular fish shape and one eye on either side of the head. By 6 months they have developed into flatfish with both eyes together on the upper side.

very small, ⅒ inch (2.5 mm) long, and helpless. They then have a normal fish shape with an eye on each side of the head. They also have a distinct air bladder, which they do not lose until they have changed into little flatfish. During this change the right eye migrates to the left side of

TURKEY

Turkeys spend most of their time on the ground, although they are strong fliers and roost in the trees at night.

upper tail coverts. The male, sometimes the female also, has a tuft of bristles near the base of the neck. The reddish legs and feet are strong, and there is a spur, as in other game birds. The head and neck are bare and are decorated with a warty bluish skin, a red throat wattle and a spurlike fleshy caruncle (outgrowth) near the base of the bill.

There are two types of wild turkey: *Meleagris gallopavo*, called simply the wild turkey, and *Agriocharis ocellata*, the ocellated turkey of Yucatan and Guatemala. There are six subspecies of *M. gallopavo*, from Pennsylvania southward. The eastern subspecies of turkey, *M. g. silvestris*, formerly ranging from Canada to Florida, was the one the Pilgrim Fathers ate at Thanksgiving. It is the most numerous wild turkey subspecies, amounting to 2.5 million birds, and the most widespread, being found in 38 states and one Canadian province. In southern Florida lives *M. g. osceola*; Merriam's wild turkey, *M. g. merriami*, lives in the foothills of the Rockies. The fourth subspecies is the Rio Grande turkey, *M. g. intermedia*. The fifth, *M. g. gallopavo*, from which present-day turkeys descended, is native to the highlands of Mexico and is the species that was domesticated by the Aztecs. Gould's wild turkey, *M. g. mexicana*, occurs in Sierra Madre Occidental from northern Chihuahua to southern Jalisco.

Easy targets

Turkeys live in open mixed woodlands, moving over the ground in small flocks by day and roosting in the trees by night. They can fly strongly but seldom stay airborne for more than a quarter of a mile (400 m). The longest turkey flight recorded was 1 mile (1.6 km). They are very shy birds and quickly disappear into the undergrowth on being disturbed. That has not saved them from hunters, however. Turkeys are vulnerable to the shotgun even at night when roosting silhouetted against a moonlit sky.

Habitat destruction

Perhaps the main cause of the decline in turkey numbers, which is particularly marked in some subspecies such as *M. g. silvestris*, has been the destruction of their habitat. The felling of trees

T HE WILD TURKEY, a native of North America, was domesticated by the Aztecs long before Columbus crossed the Atlantic Ocean. The domesticated turkey was brought to Europe and subsequently taken back to what is today the United States.

The male wild turkey is around 43 inches (110 cm) from bill tip to the end of the tail and usually weighs about 16 pounds (7.4 kg), although a weight of 22 pounds (10 kg) has been recorded. The female measures 35 inches (90 cm) and weighs around 9 pounds (4.2 kg). The plumage is mainly metallic green, bronze and copper with white or pale buff in the wings and

WILD TURKEY

CLASS	**Aves**
ORDER	**Galliformes**
FAMILY	**Meleagrididae**
GENUS AND SPECIES	***Meleagris gallopavo***

WEIGHT
(Average) Male: 16 lb. (7.4 kg); female: 9 lb. (4.2 kg); also larger specimens

LENGTH
Male: 43 in. (110 cm); female: 35 in. (90 cm)

DISTINCTIVE FEATURES
Very large; upperparts metallic green, bronze and copper; white or pale buff in wings and upper tail; bare head and neck with warty, bluish skin; red throat wattle; fleshy caruncle (outgrowth) near base of bill

DIET
Seeds, tubers, fruit, leaves, acorns; young fed on insects

BREEDING
Age at first breeding: 1 or 2 years; breeding season: April–July; number of eggs: 10 to 13; incubation period: 28 days; fledging period: young active from hatching: breeding interval: 1 year

LIFE SPAN
Not known

HABITAT
Temperate and subtropical forest to shrub-steppe; also grassland edge and agriculture edge

DISTRIBUTION
Much of United States east of the Rockies and parts of Mexico

STATUS
Not globally threatened; common in parts of range

Wild turkey

and the opening up of the land has everywhere diminished the range of wild turkeys by robbing the birds of cover as well as their natural food. However wild turkeys have been restored to most of their historic range as a result of successful management. This achievement has been brought about by trapping and relocating wild turkeys to suitable habitats, improving the habitat, increasing law enforcement and improving public support and opinion. As a result all wild turkey subspecies in the United States have increased their numbers and range.

Male wild turkeys. Outside the breeding season, males and females form separate flocks, and only the females care for the chicks.

Varied diet

Like pheasants, to which they are closely related, turkeys eat a wide variety of primarily plant foods. However insects, especially grasshoppers, are eaten in large numbers as well, as are seeds and berries, such as dewberries, blackberries and strawberries that grow in the glades and grasslands of the open forests. The turkeys also rely a good deal on acorns and nuts that have fallen from the trees and have accumulated on the forest floor and, until North America's trees were struck by fungal blight, the fruits from chestnuts. Turkeys usually drink twice a day.

Gobblers and their harems

The male turkey, or gobbler, is polygamous (has more than one mate). After mating, the females go their own ways to make their nest in a depression in the ground under low vegetation and in thickets. Each female lays one egg a day, to a total of 10 to 13. The egg is large and lightly spotted with reddish brown.

The colorful ocellated turkey is one of two forms of wild turkey. It is now confined to the Yucatan Peninsula and its numbers are low.

particularly when the chicks roost on the ground. By September half the turkey chicks have been killed.

The ocellated turkey

The ocellated turkey, which takes its name from the eyes or ocelli on its tail, has fared little better than its more familiar relative, having been wiped out in much of its former range. It now lives only on the Yucatan Peninsula in Mexico and Guatemala, where it is uncommon or rare.

It is smaller than the wild turkey, a male weighing 11 pounds (5 kg) and a female just over half this, and is more colorful than the North American species. It lacks the beard of bristles on the upper breast, and its neck is blue with red caruncles on the head. Its breeding habits are much the same except that the males and females form mixed flocks outside the breeding season, whereas in the wild turkey they separate. Ocellated turkeys also fly more freely when danger threatens, instead of relying on their legs as the wild turkey does.

Turkish merchants

The earliest known record of the domesticated turkey in Europe states that a certain Pedro Nino brought some to Spain in 1500, having bought them in Venezuela for four glass beads each. The turkey is known to have reached England by 1524, and at least by 1558 it was becoming popular at banquets.

In Spain the new bird was often referred to as the Indian fowl, an allusion that is repeated in the French *dindon*, formed from *d'Inde*. The origin of the name turkey is less obvious. One view is that it is from the bird's call *turk-turk-turk*. A more likely explanation is that in the 16th century, merchants trading along the seaboards of the Mediterranean and eastern Atlantic were known as Turks. They probably included the birds in their merchandise and these then became known as turkey fowls.

Incubation begins when the clutch is complete and lasts 28 days. The eggs usually hatch in the afternoon, the chicks being led away by the hen, with no help from the male. For the first two weeks the chicks roost on the ground. Then they fly at night to a low branch, where they settle themselves on either side of the mother, who curves her wings over the young in order to protect them.

Infancy is the most dangerous stage of a turkey's life. Predators such as opossums, raccoons and others raid the nests, and after the eggs have hatched there are further losses,

The domesticated breeds of turkeys today range from the Norfolk turkey, known in the United States as the black, through the bronze to several breeds of white turkey, including the small white. Whereas historically people wanted large turkeys to present at the table, today's domestic ovens are much smaller, and as a result the small white has gained popularity.

TURKEY VULTURE

THE TURKEY VULTURE IS one of the New World vultures, a family of birds of prey very similar in habits and appearance to the true vultures of the Old World. The family also includes the two condor species (discussed elsewhere), the yellow-headed vulture (*Cathartes burrovianus*), the greater yellow-headed vulture (*C. melambrotus*), the black vulture (*Coragyps atratus*) and the king vulture (*Sarcorhamphus papa*). The turkey vulture ranges from southern Canada to Tierra del Fuego, including the islands of the Caribbean and the Falkland Islands. The turkey vulture is up to 32 inches (81 cm) long, may weigh almost 4½ pounds (2 kg) and has a wingspan that can exceed 6 feet 6 inches (2 m). The bird has brown upperparts and blackish body plumage. The naked red head (brownish in young birds) distinguishes the turkey vulture in flight, along with the bird's long wings.

Grounded by weather

The turkey vulture's distinctive outline in flight is a common and unwelcome sight in many parts of its range, where tales of its killing livestock and spreading disease are widespread. The bird is found in many types of habitat, from forests to deserts and the high plateaus of the Andes. In most regions the turkey vulture does not migrate, but the Andean population moves to lower levels for the winter, while turkey vultures in the dry areas of the western United States fly south. Spectacular migration is noted in Mexico in fall as birds from the northern part of the range move south. Up to 1 million birds have been logged passing through Veracruz.

Turkey vultures are excellent fliers, gliding for long distances without a beat of their large wings, or soaring in thermals. It has been calculated that turkey vultures travel at 40 miles per hour (64 km/h) when they are on migration, following the lay of the land. Turkey vultures roost in trees, sometimes in groups that crouch on their perches in the manner of chickens. The birds leave the roost only when the ground has warmed up and there are rising air currents to help them take off. When it rains, turkey vultures may remain perched all day.

An appetite for carrion

Along with their relatives of both the Old and New Worlds, turkey vultures feed mainly on carrion, which may be freshly dead or in very

A turkey vulture, Baja California. This species' northern range expanded during the 20th century, perhaps due to the increased availability of carrion because of road kills.

The characteristic bare head of the turkey vulture may be an adaptation enabling the bird to feed from deep within a carcass without fouling its plumage.

	TURKEY VULTURE
CLASS	**Aves**
ORDER	**Falconiformes**
FAMILY	**Cathartidae**
GENUS AND SPECIES	***Cathartes aura***

WEIGHT
30–70 oz. (850–2,000 g)

LENGTH
Head to tail: 25¼–32 in. (64–81 cm); wingspan: 71–78¾ in. (180–200 cm)

DISTINCTIVE FEATURES
Very large size; smallish, bare-skinned head; long wings with deeply slotted primary (wingtip) feathers; brown upperparts; silvery underwing; blackish underwing coverts and body plumage; often soars in flight

DIET
Large and small carrion; live prey rarely taken; also eggs and fruit

BREEDING
Age at first breeding: 1–2 years; breeding season: very variable according to latitude, March–June in North America; number of eggs: 2; incubation period: 38–41 days; fledging period: 70–80 days; breeding interval: 1 year

LIFE SPAN
Not known

HABITAT
Very variable: humid rain forest, savanna, forest edges, desert; to altitudes of 14,100 ft. (4,300 m)

DISTRIBUTION
Breeds from southern Canada south through United States, Mexico, Central America and South America

STATUS
Common over most of range

advanced stages of decomposition. Compared with other vultures, the turkey vulture has a small bill and prefers rotten carcasses or those already opened. The animals killed by motor traffic now provide an abundant source of food. Very rarely, turkey vultures also catch small live animals, such as mice, and they occasionally take eggs and nestlings of herons and of seabirds on the Peruvian guano islands. Other foods include rotten pumpkins and the fruit of oil palms.

A circle dance

Small groups of turkey vultures sometimes perform a distinctive dance during the early part of the breeding season. About six birds gather in a clearing and hop after each other with wings outstretched. One bird hops after a second, who chases a third, and so on until they are moving in a circle. This dance apparently precedes mating. Two white eggs are laid on the floor of a cave, in a cranny, a hollow tree or on the ground in

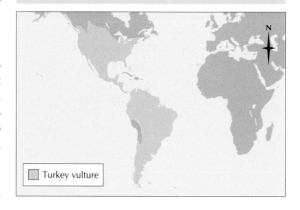

Turkey vulture

a thicket. Both parents incubate the eggs, which hatch in 38–41 days. The young are able to fly when they are about 11 weeks old.

Unjustified persecution

Humans are responsible for many turkey vulture deaths, either by accident (birds are hit by cars while feeding on animal carcasses) or through hunting. Among the reasons given for shooting turkey vultures are that they attack the nests of guano-producing seabirds and kill young livestock. The vultures' predations have no significant effect on the former, and instances of the latter must be most unusual, as the turkey vulture is too small and its bill is too weak to harm a large animal unless it is very weak or trapped. Turkey vultures are also killed, especially in the United States, because they can spread diseases such as anthrax, which they pick up on their feet and heads from carcasses. Even so, the birds are unlikely to transfer the germs to living animals. Interestingly, turkey vultures are immune to the deadly botulinus toxin, which must be a hazard to most carrion eaters.

Eyes versus nose

For more than a century there was considerable scientific controversy about the way in which turkey vultures and other birds of prey are able to locate prey or carrion from great distances. Biologists carried out many experiments that suggested turkey vultures had a very limited sense of smell and that they located their food by sight. The American ornithologist J. J. Audubon, for instance, covered a carcass with canvas and noted that turkey vultures were not attracted to the hidden meal. However, the birds did come to a canvas with a picture of a dissected sheep on it. In another experiment, a researcher concealed a putrid carcass under canvas, luring turkey vultures onto it by scattering meat on the canvas. The birds ate the visible food but were not able to detect the decaying carcass only a few inches from their nostrils.

Despite the implications of these experiments, however, scientists have recorded a number of instances of turkey vultures gathering at hidden carcasses. Turkey vultures have relatively large centers of smell in the brain, and scientists have discovered that the birds do locate their food by smell, although an animal needs to have been dead a little while before its odor is powerful enough for the birds to perceive it. The species' acute sense of smell is beneficial to certain human professions. Maintenance engineers have come to realize that turkey vultures sometimes gather over leaks in gas pipes, the odor of which apparently attracts them.

A turkey vulture on patrol over Florida. The turkey vulture belongs to one of the few genera of birds in which the sense of smell is highly developed.

TURNSTONE

Turnstones are small waders that turn stones over when they are searching for food. They are about 8½–9½ inches (22–24 cm) long, a little larger than ringed plovers. With the surfbird, *Aphriza virgata*, the two species of turnstones form a subfamily usually classed with the sandpipers but sometimes with the plovers (both discussed elsewhere). The bill is fairly short and stout; the legs also are rather short.

The turnstone, known in North America as the ruddy turnstone, *Arenaria interpres*, has a very distinctive summer plumage. The upperparts are tortoiseshell with rusty red on the back. The head, neck and underparts are white but there is a very conspicuous pattern of black lines over the head and across the breast. The legs are orange. In winter the head, breast and back become dark brown, except for the throat, which remains white.

The pattern of the turnstone's summer plumage breaks up the outline of the bird and makes it extremely difficult to see when it is feeding among the pebbles on a beach. In flight, however, the piebald pattern of the turnstone shows up very well.

The ruddy turnstone breeds along the Holarctic (northern part of the Old and New Worlds) fringe of northern North America and Eurasia, and can be found around the coasts of Alaska, northern Canada, Greenland, Scandinavia and Siberia. At one time it nested on the northern coast of Germany but it has now disappeared, probably because of the warmer climate that developed during the 20th century. When it is not breeding the ruddy turnstone is distributed around the coasts of the Indian Ocean, the Red Sea, the Persian Gulf, the Pacific Ocean, South America, Africa, Australia and New Zealand.

The black turnstone, *A. melanocephala*, breeds only in Alaska, where it nests on the coast and inland. Its plumage is more uniform than that of the ruddy turnstone. It lacks the rusty color on the back, and the head and breast are mainly black with white speckles on the side of the breast. These speckles disappear in winter. The surfbird also breeds only in Alaska, nesting in mountains above the tree line. It is mottled gray and white over most of the body, becoming uniformly darker in winter. A breeding adult has a rufous patch on the shoulder blades.

Although turnstones feed along the water's edge like other waders, they also search for and eat birds' eggs and carrion such as dead seals.

RUDDY TURNSTONE

CLASS	**Aves**
ORDER	**Charadriiformes**
FAMILY	**Scolopacidae**
SUBFAMILY	**Arenariidae**
GENUS AND SPECIES	***Arenaria interpres***

WEIGHT
3–5½ oz. (85–150 g)

LENGTH
8½–9½ in. (22–24 cm)

DISTINCTIVE FEATURES
Plump, medium-sized shorebird; rufous and black upperparts; black on breast over white underparts; orange legs

DIET
Mostly insects in breeding season; at other times, invertebrates, crustaceans, mollusks and carrion such as dead fish

BREEDING
Age at first breeding: 1 year; breeding season: mid-May–early July; number of eggs: 4; incubation period: 22–24 days; fledging period: 19–21 days; breeding interval: 1 year

LIFE SPAN
Not known

HABITAT
Breeds in Holarctic (northern part of Old and New Worlds), on rock, clay or shingle; otherwise, entirely coastal, preferring stony, rocky or seaweed-strewn shores

DISTRIBUTION
Holarctic fringe of northern North America and Eurasia for breeding; otherwise, coasts around Indian Ocean, Red Sea, Persian Gulf, Pacific Ocean, South America, Africa, Australia and New Zealand

STATUS
Common

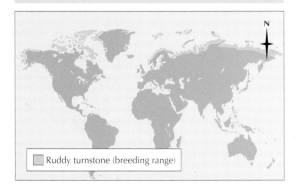

Ruddy turnstone (breeding range)

Speedy migration

Turnstones migrate long distances between their breeding grounds and wintering grounds. The ruddy turnstone, for example, travels south to the Cape Province of South Africa, Chile and even Australia and New Zealand. Some, presumably juveniles, stay there while the remainder fly north to breed. The black turnstone and the surfbird spread down the Pacific coast of the United States. One ruddy turnstone gave a good indication of the speed at which birds migrate. It was ringed in Germany at 11 a.m. one day and was recovered on the northwestern coast of France at 12 a.m. the following day, having traveled just over 500 miles at an average of 38 miles per hour (61 km per hour).

While they are on migration, turnstones fly in large flocks at a considerable height, but when they are not traveling, they fly low and in small parties, circling over the sea and returning quickly to the shore when disturbed. Turnstones often associate with other waders and fight them when competing for food.

Derivation of the name

When they feed along the shore, turnstones eat small animals such as winkles, worms and crustaceans, as well as the eggs of other birds and carrion. Inland on the breeding grounds they feed mainly on insects and their larvae, such as

Its slightly upturned bill enables the turnstone to flick over pebbles and seaweed to search for small crustaceans, mollusks and dead fish.

beetles and caterpillars. Their name is derived from the way they flick over stones, clods of earth or seaweed to expose the small animals underneath. They bend their legs, then insert the slightly up-curved bill under the object and flick it over, sometimes pushing with the breast if it gets stuck halfway. Stones up to ½ pound (230 g) can be shifted in this way, flat ones being more difficult to move than round ones.

It sometimes appears as though turnstones cooperate with each other to turn heavy stones, but they are just as likely to work against each other as to lever in the same direction. Sandhoppers are found by flicking over seaweed, whereas small mollusks and worms are exposed by digging into the sand. The stone-turning habit, however, is not so common as the name suggests, because turnstones spend much of their time feeding along the water's edge, a characteristic typical of many other waders.

Inland nest sites

Ruddy turnstones usually nest near the shore, but sometimes they choose islands some distance up rivers. The black turnstone and the surfbird regularly nest inland, the former on the banks of pools in the tundra and the latter in rocky outcrops in mountainous regions.

Nonbreeding turnstones lack the rufous color on their upperparts, and fly south to warmer climates.

The nest is a scantily lined scrape in the ground in which four eggs are laid. Both parents incubate the eggs and feed the chicks. Turnstones defend their broods vigorously, harassing any predators, especially skuas and Arctic foxes.

Nest raiders

Turning over stones and clods is not the turnstones' only feeding habit. Biologist Alexander Wetmore observed the way in which migrating turnstones feed on terns' eggs on Laysan Island, Hawaii. Whenever the terns were disturbed and flew off their nests, parties of turnstones would alight and peck holes in the eggs. The turnstones also tried to attack the eggs of boobies and frigate birds but these were too tough. Two turnstones even dragged an egg from under a sitting tern and immediately ate it.

Turnstones have also been reported to eat carrion. In Alaska, for example, turnstones seem to feed on the carcasses of slaughtered fur seals, although it is possible they may be feeding only on the maggots in the carcasses. However, in *British Birds* (1966) another observer recorded seeing turnstones definitely feeding on the flesh of a human corpse washed up on the beach. Turnstones also feed on human refuse, such as potato peelings.

TURTLEDOVE

THE TURTLEDOVE IS ONE of the smaller members of the pigeon family and one of the least harmful to agriculture. Its habits, especially its migrations, are particularly closely linked to its food supply. It is about 10¼–11 inches (26–28 cm) long, light chestnut on the upperparts, with delicate black markings, gray on the head, and pinkish buff underparts. The centers of the wing feathers are black. The eye is deep red. The bill is relatively weak. There are patches of black-and-white feathers on the sides of the neck, used in display. The fan-shaped tail has long, white-tipped black feathers. The turtledove winters in tropical Africa, south of the Sahara but north of the equator, and migrates north to breed in northern Africa, southwestern Asia and Europe.

In contrast to the wood pigeon, with which it shares a similar summer range, the turtledove favors open woodlands, parkland and shrubberies. It is seldom found on buildings and prefers perching on low or medium-sized trees and bushes near cultivated ground. The turtledove has a purring call, from which its Latin name *turtur* comes. Originally called dove in Old English, it later became the turtur dove and then the turtledove.

Feeds on weed seeds

The summer diet of turtledoves has been comprehensively analyzed by scientists. From the end of April or the beginning of May, when the doves reach their summer grounds, to the end of June, they feed mainly in hay fields. From July to September they feed mainly among cereal and other crops. Their food is 95 percent seeds of weeds, with about 3 percent animal food, mainly small snails. The weed seeds include those of grasses but are predominantly the fruits of fumitory *Fumaria* spp. The seeds of millet, a commercial crop in the former Soviet Union, are particularly suitable for the turtledove, and in that region it is regarded as a potential pest.

Elsewhere, the distribution of turtledoves is closely linked with that of the fumitory. In Britain, for example, the fumitory, a weed that grows on cultivated ground, is common over much of England, rarer in Wales and is absent from most of Scotland. The distribution of the turtledove follows these limits. Its range has contracted in northwest Europe in recent years.

Bobbing courtship

The territorial display of the turtledove consists of launching out from a perch, then climbing steeply before gliding down and circling back to the perch again. During courtship the male displays to the female by puffing out his chest and bobbing up and down with a lowered bill.

The nest is a frail platform of thin twigs, sometimes lined with roots, hair or plant stems. One or two white, oval or elliptical eggs are laid, and usually each pair of doves has two or three clutches per season. The parents share the incubation for 13–14 days and feed the nestlings on pigeon's milk, a secretion from their crop, for 18 days. Egg-laying reaches its peak in the second half of May and early June, declining during July.

The supply of seeds improves as the summer wears on, and there is a higher rate of survival among young born in the second half of the breeding season. This is because the parents need to spend less time looking for food, so the losses from nest-robbing predators are fewer. Even so, there are heavy losses over the season as a whole. Nest-robbers take 34 percent of the eggs, only 47 percent hatch and only 39 percent produce fledged young. The nest-robbers are other birds, such as the magpie and the jay.

Learning the hard way

The last egg-laying tends to be in early July. As the season progresses, the breeding impulse weakens and the parents may desert both eggs and young. Even if they continue to feed the nestlings, their waning parental instinct means the young may not get sufficient food to lay in the stores of fat needed on migration. This leads to further deaths on migration, already perilous for young doves. Ornithologists have found that on the journey south through western Europe in

The turtledove is a summer visitor over most of Europe and some parts of North Africa and Asia.

The turtledove feeds its chicks on pigeon's milk, a nutritious, curdlike liquid that is regurgitated from the crop of the adult bird.

September, the young and inexperienced birds take a more westerly route than the old birds, flying down the coast of Portugal. There is danger on this route of their being blown out to sea by the prevalent easterly winds. Moreover, young birds tend to migrate under unfavorable weather conditions, so instead of being able to fly high, they must come lower, within range of the shotgun. On Malta, for example, up to 20,000 turtledoves a day are shot at the peak spring passage, with around 100,000 shot annually.

Hawks and egg robbers

The predators of turtledoves are mainly hawks, such as sparrowhawks, and the nest-robbers, including the magpie and jay. The latter are always on the lookout for unguarded nests, so their impact is greatest during the early part of the season, when food is less plentiful and the parents are away from the nest more often. Overall there is a 50 percent mortality among adult turtledoves, made good each year by surviving young ones.

Migratory molt

In summer, turtledoves work to a tight schedule as breeding continues late into the year, leaving little time before migration to go through a molt. This may be because they have insufficient energy to spare for this exacting process without depriving the body of fat reserves needed for migration. As a result, they start the molt and then migrate with an arrested molt, which is completed when they reach their winter quarters.

The question arises as to why turtledoves should migrate at all. Climate seems not to be the deciding factor because turtledoves can be kept in outdoor aviaries during the winter in England, if they are well fed. In the wild there should be no shortage of weed seeds on the ground.

TURTLEDOVE

CLASS	**Aves**
ORDER	**Columbiformes**
FAMILY	**Columbidae**
GENUS AND SPECIES	***Streptopelia turtur***

WEIGHT
3½–6¼ oz. (100–180 g)

LENGTH
10¼–11 in. (26–28 cm)

DISTINCTIVE FEATURES
Light chestnut upperparts with black markings; pinkish-buff underparts; black-and-white rimmed, fan-shaped tail; black-and-white marks on neck sides

DIET
Mostly seeds and fruits of weeds and cereals

BREEDING
Age at first breeding: 1 year; breeding season: May–August; number of eggs: 1 or 2; incubation period: 13–14 days; fledging period: 20 days; breeding interval: 1 year

LIFE SPAN
Not known

HABITAT
Open woodlands, parkland and shrubberies

DISTRIBUTION
Breeds from Iberia and southern England across Europe to southwest Asia and parts of north Africa; winters in sub-Saharan Africa

STATUS
Common or fairly common

Eurasian turtledove (breeding range)

The need to migrate may well be linked with the pattern of their feeding behavior. They take mainly the seeds from standing vegetation. The fumitory, and other plants supplying their food, die down at summer's end, and turtledoves have little skill in taking seeds from the ground.

TURTLES

Turtles are reptiles that have a shell protecting most of their body. The shell is made up of two parts: the upper part is called the carapace and the lower part is called the plastron. The carapace and the plastron are joined along the sides of the body, but at the front end there is an opening for the head and legs to emerge, and at the back end an opening for the back legs and tail. Most turtles are able to withdraw their head, legs and tail into the shell so that it is almost impossible for a predator to kill and eat them unless it is large enough to swallow them whole. Mud turtles, genus *Kinosternon*, and musk turtles, genus *Sternotherus*, are notable in that their shells have hinged lobes at the front and back. When a mud or musk turtle withdraws into its shell, the lobes can be pulled shut, enabling the turtle to conceal itself completely.

The smallest turtle species is the speckled tortoise, *Homopus signatus*, of South Africa and Namibia, the largest specimens of which are about 3¾ inches (9.5 cm) in length and weigh just 5 ounces (140 g). At the other extreme, leatherback or leathery turtles, *Dermochelys coriacea*, can grow to 9 feet (2.7 m) in length and weigh more than ¾ ton (860 kg).

Classification

Their protective shells are one of the factors that have made turtles such successful animals in evolutionary terms. They are the oldest form of reptile alive today and have survived with very little change for more than 200 million years. Scientists have dated the oldest turtle fossils back to the Triassic period, about 230 million years ago.

Concerns for the survival of the gopher tortoise, Gopherus polyphemus, *of the southern United States, have resulted in full legal protection for the species, the control of illegal harvests and the relocation of vulnerable populations.*

There are two major groups of turtles, classified according to the way in which the neck is bent when the head is withdrawn into the shell. In the main group (the hidden-necked turtles, suborder Cryptodira), the neck is bent on a vertical plane, in a similar manner to the neck of a cobra. In the side-necked turtles (suborder Pleurodira) it is folded sideways beneath the front edge of the upper shell, that is, on a horizontal plane.

Side-necked turtles usually have longer necks than hidden-necked turtles; in some species, the neck is so long that they are called snake-necked turtles (family Chelidae). The two suborders of turtles also differ from one another in various key aspects of their skull and skeleton structure.

CLASSIFICATION
CLASS Reptilia
ORDER Testudines
SUBORDER Cryptodira: hidden-necked turtles; Pleurodira: side-necked turtles
FAMILY 12
NUMBER OF SPECIES About 250

The shell of a turtle has a relatively complicated structure. The interior is made up of bony plates that are fused together. These bony plates are part of the skeleton, a unique feature among reptiles. Some of the bones are derived from the backbone, others from the ribs. The outside of the shell is composed of horny plates, which are often known as shields or scutes. However, the joins between individual scutes do not correspond to the joins between the underlying bones. Three kinds of turtles do not have scutes, but instead have the bones covered in a layer of soft, leatherlike skin. These are the soft-shelled turtles (family Trionychidae); the Papuan, pig-nosed or hog-nosed turtle (*Carettochelys insculpta*); and the leatherback or leathery turtle.

Found in warm regions

Turtles are found in most of the warmer parts of the world, on land, in freshwater rivers, swamps and lakes, and in the sea. The suborder of hidden-necked turtles includes the majority of species that live on land, in rivers and in lakes, and also contains the species that inhabit tropical and temperate oceans. Side-necked turtles are found exclusively in fresh water and occur only in central and southern Africa, Australasia and South America. There are no turtles living in the Arabian Peninsula or in South America west of the Andes Mountains, and they are also absent from the very arid deserts of central Australia and from New Zealand.

Turtles, tortoises and terrapins

The scientific group in which turtles are placed has been given several names in the past. At one time it was called the order Chelonia, and as a result turtles and their relatives are some-

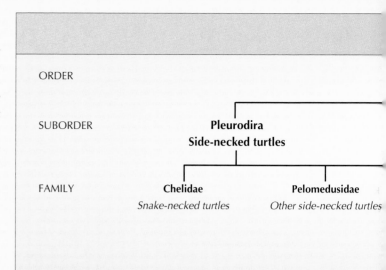

times called chelonians. To a certain extent, the differences between turtles, terrapins and tortoises depend on where in the world they are found. In North America, most chelonians are called turtles, although species that never go in the water are often called tortoises. Small chelonians that live in rivers, ponds and streams are usually also called turtles, but some also have their own individual names. These include terrapins, cooters and sliders. The first two of these terms are derived from local Native American names, while the last arose because the animals appear to slide into the water after they have been basking on land in the sunshine. In England, the terms are used differently. Chelonians that live on the land are always called tortoises, those in the sea or in lakes and rivers are called turtles and those in ponds and streams are called terrapins.

Understandably, such uncertain classification often leads to confusion. Part of the problem arises from the fact that the names do not reflect any scientific differences; there is no strictly technical definition of a turtle, a tortoise or a terrapin. For example, in the woods of the southern Appalachian Mountains, a hiker may come across chelonians that live on land. By conventional scientific classification, these ought to be categorized as tortoises. In fact, the animals are called box turtles, and they belong to the genus *Terrapene*.

No teeth

In addition to the shell, all of the species of modern turtles have a number of other features in common. One is that

The endangered Pacific, or olive, ridley turtle (Lepidochelys olivacea) is a sea turtle known for its mass nestings, which occur at only five beaches in the world. More than 300 females may come ashore to lay their eggs at one time.

Turtles Family Tree

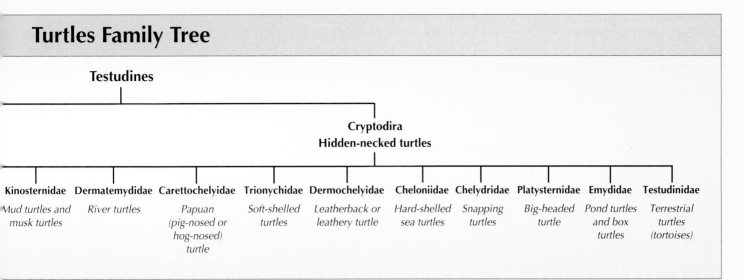

Testudines

Cryptodira
Hidden-necked turtles

Kinosternidae	Dermatemydidae	Carettochelyidae	Trionychidae	Dermochelyidae	Cheloniidae	Chelydridae	Platysternidae	Emydidae	Testudinidae
Mud turtles and musk turtles	*River turtles*	*Papuan (pig-nosed or hog-nosed) turtle*	*Soft-shelled turtles*	*Leatherback or leathery turtle*	*Hard-shelled sea turtles*	*Snapping turtles*	*Big-headed turtle*	*Pond turtles and box turtles*	*Terrestrial turtles (tortoises)*

they have no teeth. The edges of the jaws are made of very hard skin, and this can be surprisingly sharp. An alligator snapper, *Macroclemys temmincki*, which is the largest freshwater turtle in North America, can easily sever a human finger or toe with one powerful bite. This has been a feature of turtles since very early in their evolution, although many fossil forms have teeth in the palette of the mouth. They also have massive, strong skulls. Finally, all turtles lay eggs, which are always placed in a burrow on land. Species that live mostly in the water come ashore to lay their eggs. A female sea turtle may live for 30–50 years, but only a few days of that time are spent on land. As an adult the female comes out on to a beach to lay her eggs, and the young turtle will have to scuttle over the beach to the sea when it hatches. Males spend only a few minutes on land, when they first hatch. Courtship and mating take place in the water, and so after hatching a male never emerges from the sea again.

Mass nestings

Nesting by sea turtles and some of the bigger species of turtles found in rivers is communal. Many hundreds of females may emerge at the same time, almost always by night. Female loggerhead turtles, *Caretta caretta*, emerge from the sea at night to dig a shallow hole in the sand some distance above the high-tide line. They then deposit a clutch of about 120 eggs, cover it over with sand and return to the sea. The procedure may take several hours to complete and is very tiring for the turtles. Mass nestings are a spectacular sight, and have become a tourist attraction in many places. Females return year after year to the same nesting sites in order to lay their eggs. However, this makes the animals very vulnerable to hunters, who kill them for their meat and for the horny plates of their shells. The Atlantic ridley turtle, *Lepidochelys kempii*, apparently uses only one nesting beach, located on the eastern coast of Mexico, making it particularly open to overexploitation.

A growing number of turtle species is now in danger of extinction. Many eggs never survive to hatch because they are taken by birds or mammals. When the eggs hatch, the hatchlings emerge and have to make their own way into the sea without assistance or protection. The most vulnerable species are the western swamp turtle, *Pseudemydura umbrina*, which

Sea turtles (green turtle, above) are at their most vulnerable just after they have hatched and are making their way to the sea. At this time they are attacked by seabirds and crabs.

lives in a few small swamps near Perth in the state of Western Australia, and the black soft-shelled turtle, *Aspideretes nigricans*, which now lives only in a large pond at the Sultan Bagu Bastan shrine near Chittagong in Bangladesh.

Diverse lifestyles

The fact that turtles have a shell has placed limits on the amount of variation that has occurred in their bodies, but in other respects they are a surprisingly diverse group. One of the areas in which there is much variation is their diet. Even among the eight species of large sea turtles, there are great differences. Loggerhead turtles eat mainly sea snails, mussels and crabs. Hawksbill turtles, *Eretmochelys imbricata*, are more specialized, feeding mostly on sponges that they scrape off the surface of rocks, although they also take crabs, fish, aquatic plants and jellyfish. Hawksbills, so called because of their hooked beak that resembles a hawk's bill, appear to be immune to the poison from the Portuguese man-of-war, *Physalia physalis*, which forms part of their diet. Leatherback turtles wander the seas searching for jellyfish. Green turtles, *Chelonia mydas*, feed primarily on turtle grass and seaweed.

Most of the turtles that live on land are vegetarian, although some species also eat small insects and other invertebrates. Many of the land turtles live in deserts and other arid

Africa has more species of land turtles than anywhere else on earth. Pictured is the leopard tortoise, Geochelone pardalis, which is widespread in the southeast of the continent.

places, and they may spend a lot of time roaming widely in search of the juiciest leaves and flowers. Turtles in freshwater habitats usually feed mostly on insects and snails, but many of them eat vegetation as well. Some of the bigger species feed predominantly on fish. The alligator snapping turtle from the southeast of the United States has a special knob on its tongue. The turtle lies on the bottom of a river with its mouth open and wriggles the knob. The latter acts as a lure for unsuspecting fish, which take it to be a juicy worm.

Most individual features of the structure and biology of animals have both advantages and disadvantages, and the shell of turtles is no exception. It has an obvious role in providing protection for the animals, but on the other hand it also makes them heavy and cumbersome and makes breathing difficult. In water, the weight of the shell is effectively reduced by the buoyancy of the body, but on land it must be fully supported by the legs. Most land turtles are slow, ponderous animals, as as are water turtles when they emerge onto the land. However, when they are swimming, flapping the front legs, which are enlarged and flattened as flippers, they are graceful.

Temperature control

Turtles, like all reptiles, cannot control their body temperature using heat produced chemically from food in their bodies. Often they are warm to the touch, however. This is because they have been in the heat of the sun, or have placed themselves on rocks or soil that have in turn been warmed up by the sun's rays. This process is called behavioral thermoregula-

Many of the land turtles live in deserts and other arid places. Ranging across southern and eastern Africa, the Cape terrapin, Pelomedusa subrufa, is also found in the Yemen, making it the only side-necked turtle with an Asian distribution.

tion. It is most effective for turtles on land, which will often seek out warm places, especially in the morning. Desert tortoises, *Gopherus agassizii*, of the Mojave Desert in Utah, Nevada and California, dig long burrows into the sandy soil. They come to the entrances of their burrows in the morning and bask there in the sun until they have warmed up. Only then do they set off in search of food.

Turtles in freshwater habitats often come out from the water to bask in the sun, frequently on a nearby rock or branch so that they can escape to the water again if they sense danger in the locality. Small freshwater turtles are particularly common in the northeast of the United States and in southern Canada, where the sun is often weak and the air temperature cool, so they spend a lot of time basking. However, they also frequently emerge from the water when the sun is not shining. Scientists are uncertain as to the reason for this. It may be that such movement discourages leeches.

Turtles in the sea often spend long periods at the surface, which may help them keep warm if the sun is shining. Some species also have the blood vessels in their flippers arranged so that much of the heat produced from their swimming muscles is carried back into the body. In this way they may keep their body temperature higher than that of their surroundings.

Endangered giants

The largest species of turtle alive today is the leatherback turtle, which roams the warmer seas of the world searching for jellyfish. The biggest land species is the Aldabra tortoise, *Aldabrachelys elephantina*, which is native to the small island of Aldabra and a few other islands in the Indian Ocean. The shell of the largest specimen ever reliably recorded was 54 inches (1.4 m) in length and the animal weighed more than 530 pounds (250 kg). Almost as large are the giant tortoises, *Geochelone elephantopus*, native to the Galapagos Islands in the Pacific Ocean, off the coast of Ecuador. There are several subspecies of giant tortoises, each confined to a separate island or group of islands. They have no natural predators as adults but are vulnerable to habitat loss and destruction, especially by goats, and their eggs are eaten by a wide range of predators including dogs, pigs and rats. The larger bones of both Aldabran and Galapagos giant tortoises have a distinctive honeycomb structure to help to reduce their weight. The shells of both species are similar in structure, containing many small air chambers that make it possible for the animals to bear their considerable weight.

During the 19th century, the numbers of Galapagos giant tortoises were reduced considerably by whaling and sealing ships. The crews of these ships took the still-live animals on board to provide a source of fresh meat at sea, because the tortoises could live for up to 1 year in the ships' holds, without food or water. On the island of Pinta, to the north of the Galapagos Archipelago, goats introduced by local fishers in the 1950s destroyed much of the vegetation, leaving the giant tortoises' nests more exposed to predators. The goats also competed with the tortoises for food. In 1971, National Park wardens discovered a single male giant tortoise alive on Pinta. They named him Lonesome George and brought him back to become part of the captive breeding program at the Charles Darwin Research Station. However, despite being paired with females that seemed genetically close to him, no eggs have so far been produced. Scientists have discussed cloning the tortoise and manipulating the clone's gender to produce a female but regard this very much as a final option. If research teams fail to find other Galapagos giant tortoises on Pinta and further attempts to produce offspring prove fruitless, Lonesome George's race will end with his death, as has been the case already for several other subspecies of Galapagos giant tortoises.

For particular species see:
- BIG-HEADED TURTLE • GREEN TURTLE
- HAWKSBILL TURTLE • HIDDEN-NECKED TURTLE
- LEATHERBACK TURTLE • LOGGERHEAD TURTLE
- MUD TURTLE • PAINTED TURTLE • PAPUAN TURTLE
- SNAKE-NECKED TURTLE • SNAPPING TURTLE
- SOFT-SHELLED TURTLE • TERRAPIN • TORTOISE

TYRANT FLYCATCHER

An eastern kingbird, New Hampshire. Kingbirds are also called bee martins, although they prefer to eat other insects.

THE TYRANT FLYCATCHERS ARE a vast group of birds, with around 425 species including birds with the common names of kingbird, kiskadee, phoebe and pewee. They are similar but not related to the true flycatchers (discussed elsewhere). Confined to the New World, tyrant flycatchers have taken up many ways of life and are so widely varied in appearance that generalization is nearly impossible. Many have a short crest and a bright patch of color on the crown. The wings may be rounded or pointed and the bills also vary greatly. Tyrant flycatchers range in length from 3–15 inches (7.5–37.5 cm).

Some of the tyrant flycatchers are brightly colored. The vermilion flycatcher, *Pyrocephalus rubinus*, has a red head and underparts with a brown back. The many-colored tyrant, *Tachuris rubrigastra*, has feathers of black, blue, green, orange, scarlet and white. The 15-inch (37.5-cm) scissor-tailed flycatcher, *Muscivora forficata*, has a black-and-white forked tail making up over half its total length, and the northern royal flycatcher, *Onychorhynchus mexicanus*, has a fan-shaped crest of orange and violet. Most tyrant flycatchers are, however, rather inconspicuously colored with shades of brown and gray.

Tyrant flycatchers range from the tree line in Canada to the tip of South America, including the Galapagos and Falkland Islands, but most are found in the Tropics.

Variable forms

The habitats of tyrant flycatchers are extremely varied and this is directly related to the variability in their physical form. Long-winged species are migratory, whereas those with short, rounded wings live in forests, and in open country there are tyrant flycatchers with strong legs, similar to those of pipits. Many tyrant flycatchers are found near water, for example the eastern phoebe, or water pewee, *Sayornis fusca*, which is a gray-colored bird named for its plaintive, repetitive call. It is identifiable by its habit of flicking or bobbing its tail while perched.

Tyrant flycatchers range from sea level to 12,000 feet (3,600 m) in the Andes and live in grassland, dense forests and swamps. Not surprisingly, the feeding and nesting habits of tyrant flycatchers are also very varied. Their songs are poor and not particularly melodic. The kiskadees are also named after their calls, just as the pewees are.

Many feeding habits

In their manner of feeding, the tyrant flycatchers demonstrate many of the habits familiar from Old World flycatchers. Most feed on insects, but their diet is often supplemented with fruit and larger animals. In common with the Old World flycatchers, they wait on a favored perch and fly out to catch passing insects with an audible snap of the bill. Their ability to catch flying insects is assisted by the rictal bristles around the base of the bill, which act as a net. Rictal bristles are best developed in the fly-catching species and are greatly reduced in those species that catch larger prey or eat fruit.

Instead of catching flying insects, some tyrant flycatchers, such as the gray monjita, pounce on terrestrial insects, as shrikes do, and those with strong legs chase them on the ground, leaping up to snap a low-flying insect. Others hunt insects among foliage. The larger species, such as the boat-billed flycatcher, *Megarhynchus pitangua*, feed on small birds, lizards, frogs and mice, battering their victims before tearing them apart. The black phoebe, *Sayornis nigricans*, catches small fish, and to complete this example of contrasts, the cattle tyrant searches the backs of cattle for insects and ticks, in a similar manner to an oxpecker.

VERMILION FLYCATCHER

CLASS	**Aves**
ORDER	**Passeriformes**
FAMILY	**Tyrannidae**
GENUS AND SPECIES	***Pyrocephalus rubinus***

WEIGHT
½ oz. (14.5 g)

LENGTH
Head to tail: 6 in. (15 cm)

DISTINCTIVE FEATURES
Small and short tailed. Male: brilliant scarlet head and underparts; brown back; black bill, eyestripe, wings and upperparts. Female: mid-gray upperparts; gray mask; streaked white breast; pinkish vent and undertail coverts

DIET
Insects

BREEDING
Age at first breeding: 1 year; breeding season: April–June; number of eggs: 2 or 3; incubation period: 14–15 days; fledging period: 14–16 days; breeding interval: 1 year

LIFE SPAN
Not known

HABITAT
Open trees and bushes near water

DISTRIBUTION
Southwest U.S. and Mexico

STATUS
Common

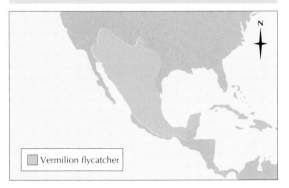

Vermilion flycatcher

In all but a few tyrant flycatchers, the sexes are similarly marked. One notable exception is the vermilion flycatcher, the male of which is fiery red with dark wings and back.

Snakeskin decorations

Tyrant flycatchers are almost as varied in their nesting habits as they are in their feeding habits. The nests may be open cups or domed nests. They may be located on the ground or in the fork of a tree and they may be either camouflaged or conspicuous. Some tyrant flycatchers take over the holes of woodpeckers, whereas others use the mud nests of ovenbirds and a few nest near wasps, sheltering under their protection. Most species, however, build cup-shaped nests of grass and twigs in the branches of trees. Many tropical species build purselike domed nests that have tubular entrances. Most tyrant flycatchers are territorial during the breeding season, and attack other species of birds that venture too close during this period.

The great-crested flycatcher, a hole-nester that takes over woodpecker holes, birdhouses and similar sites, sometimes decorates its nest with cast-off snakeskins. Presumably, the snakeskins are merely one of the many objects that have been found in birds' nests that have been taken simply because they are convenient materials located near at hand.

The eggs of tyrant flycatchers vary in coloration from pure white to white with spots and streaks, and the clutch number varies from two in the Tropics to four in higher latitudes. Both parents construct the nest, but only the female incubates the eggs. The incubation and fledging periods also vary. They usually are 14–18 days and 13–14 days, respectively, but incubation may last 19–3 days and the fledging period may be 21–25 days. Both parents feed the young, and there may be two or more broods a year, especially in the Tropics.

Little tyrants

The kingbirds of the genus *Tyrannus* are aptly named because of the manner in which they harass other birds that come into their territories. The eastern kingbird of North America, *T. tyrannus*, was once known as the little chief due to the way that it attacked humans who ventured near its nest, and did not hesitate to attack crows or hawks considerably larger than itself. Even if the intruders show no interest in the kingbird and its brood, they are pursued until they are well away from the nest.

Kingbirds are not the only aggressive species of flycatchers, however. One ornithologist has described seeing an acadian flycatcher, *Empidonax virescens*, attack cowbirds, which may attempt to lay their eggs in its nest, squirrels, chipmunks and even inoffensive nuthatches.

Tyrant flycatchers usually have rounded or shallowly forked tails but the scissor-tailed flycatcher has a very long, deeply forked tail, and can perform aerial acrobatics.

UAKARI

THE UAKARI (pronounced wakari) is a South American monkey, closely related to the saki (discussed elsewhere) but differing from it in a number of features, especially the short tail. It is the only South American monkey in which the tail, measuring only 4¾–8¾ inches (12–22 cm), is shorter than the head and body, at 12–24 inches (30–60 cm). The body is covered with shaggy hair, variable in color, but underneath it the uakari is a skinny, spidery animal. The top of the head and the face are naked, with only very short, sparse hair or none at all. This baldness is accentuated by an almost complete lack of fat under the skin, making the face very lean and bony. In the adult male, the jaw muscles become big and bulky and can be seen clearly beneath the bare skin, bulging out on the top of the skull. The bare face pokes out from a mass of shaggy hair that begins behind the ears and on the neck and the back of the head.

There are two species of uakari, found around the Amazon on both sides of its upper course. On the north side, between the river Branco and its tributary the Rio Negro, and Japurá, lives the black-headed uakari, *Cacajao melanocephalus*, which is chestnut brown in color with black hands and feet and a naked black face. To the southwest, on the other side of the Rio Japurá, lives the bald uakari, *C. calvus*, which differs in its skull characteristics and has a longer coat and a pink or red face. The face turns pale if the animal is kept from sunlight and becomes bright crimson if it is allowed to live in the full sun. The bald uakari is divided into two distinct races: the white one, in which the coat is white or silvery, and the red one, with a coat that is red like the face. The first ranges from the Rio Japurá to the Rio Içá. The second extends south from the Içá to about 7° S, its range bounded on the west and east by the rivers Ucayali and Juruá.

Active in the treetops

Uakaris have rarely been observed in the wild. Their thick, shaggy coats once led people to suspect they were clumsy and lethargic animals. However, beneath their fur uakaris are slim, and are very agile and active. In the wild they have been seen making leaps of 20 feet (6 m) or so, launching themselves into the air with arms stretched forward. On the ground they are somewhat ill at ease, walking with the hands partly flexed and turned out sideways. In captivity, however, scientists have observed them inventing games for themselves, sliding along the cage floor or turning back somersaults.

Uakaris are highly dexterous when feeding themselves. They hold the food in the whole flexed hand, the thumb not being divergent, or between the index and middle fingers, or even between the hand and wrist. They have projecting lower incisors that are probably used for spearing the fruits that form part of their diet along with buds, leaves and seeds.

Uakaris have been seen both in small troops and in large gatherings of about 100. They may go right up to the treetops but they travel through the forest among the lower branches.

Sedation an old trick

One of the first authentic records of uakaris in their natural habitat was written nearly 150 years ago. In 1855 the writer observed that the bald

Female white bald uakari, Amazonas State, north of Brazil. Quick and agile in the upper forest canopy, the uakari is much less nimble on the ground.

Uakaris can use their hands very skillfully when feeding, even though the thumbs are nondivergent. Pictured is a white uakari.

UAKARIS

CLASS	**Mammalia**
ORDER	**Primates**
FAMILY	**Cebidae**

GENUS AND SPECIES **Bald uakari, *Cacajao calvus*; black-headed uakari, *C. melanocephalus***

ALTERNATIVE NAME
Uacari

WEIGHT
5½–7¾ lb. (2.5–3.5 kg)

LENGTH
Head and body: 12–24 in. (30–60 cm); tail: 4¾–8¾ in. (12–22 cm)

DISTINCTIVE FEATURES
Short tail, less than half body length; nearly naked face; large ears; shaggy fur; variable color among subspecies, white, red, brown or black

DIET
Fruit, leaves, insects, eggs and small animals

BREEDING
Age at first breeding: female 3 years, male slightly older; breeding season: May–October; number of young: 1; gestation period: about 6 months; breeding interval: 2 years

LIFE SPAN
Up to about 32 years in captivity

HABITAT
Forest trees, usually around lakes or small rivers

DISTRIBUTION
Around upper course of Amazon River, northern South America

STATUS
Both species endangered

Uakaris

uakari was captured alive by shooting it with arrows and blowpipes. The curare poison used with these arrows was diluted to capture the monkeys alive. After it had been shot, a uakari would run quite a long way before it fell from the branches, weak with the poison. As soon as the animal fell, a pinch of salt was put into its mouth, which, so the story ran, acted as an antidote to the poison and revived the monkey. Animals caught in this way were kept as pets, often being traded far from their native haunts. They seem to have developed a great devotion to their owners, and they were fairly easy to keep, although the initial death rate was high. Today they are not too uncommon in zoos, and some have lived as long as 32 years in captivity. Their appearance, especially in the case of the bald species, and their offbeat antics, performed in silence, have made them popular zoo inmates.

The method employed by South Americans in the past to capture live uakaris anticipated modern methods. Today, animals marked for study purposes or for transport to wildlife parks are first immobilized with a drug in a dart shot from a crossbow, a similar technique to the blowpipe darts used long ago by South Americans.

URANIA MOTH

THE NAME URANIA WAS first given to New World moths, but it has been extended to include all moths of the family Uraniidae. Urania moths are almost entirely limited to the Tropics. Some are large, conspicuous day-flying moths, brilliantly colored and bearing tail-like extensions on the hind wings similar to those of the swallowtail butterflies (discussed elsewhere), which these urania moths greatly resemble in appearance. Other uranias are duller, pale, lack tails on their wings and fly by night.

Mainly tropical distribution

The family Uraniidae contains several genera. The genus *Urania* is centered in tropical America and most of its members are brilliantly colored. The South American *U. leilus* is iridescent green and blue with a long tail and broad white fringes. It has a slender body and a wingspan of 3 inches (7.6 cm). The largest and most brilliant member of the family belongs to the genus *Chrysiridia*. The Madagascan sunset moth (*C. riphearia*) has been called the most magnificently

colored of all animals. Found only on the island of Madagascar in the Indian Ocean to the east of Africa, its wings are black and banded with metallic green, which changes to blue and gold according to the angle of the light. Its hind wings each have a patch of glowing copper and purple, together with long white fringes around their tailed and deeply scalloped margins.

In the region extending from India to Australia, the genera *Nyctalemon* and *Alcides* are similarly swallow-tailed and some are beautifully colored with bands of pale blue, green and yellow. Australia's zodiac moth (*Alcides zodiaca*) occurs in rain forest in the tropical north of the country and may be seen anywhere from ground level to the tops of trees. The moth has a wingspan of 4 inches (10 cm) and is black above with areas of iridescent yellow and pink. Below, it is iridescent pale green banded with black. The common *Nyctalemon patroclus* of the region is more soberly colored, being banded with brown and white, and frequently perches on trees and buildings with its wings outspread.

The uraniid moth **U. leilus** *basks on the ground in Rio Madre de Dios, Peru. Butterflies usually close their wings at rest, but moths tend to sit with outspread wings.*

Less brilliantly colored than some uraniids, Nyctalemon patroclus has a wingspan of about 5 inches (13 cm) and ranges from India as far as the tropical north of Australia.

URANIA MOTHS

PHYLUM	**Arthropoda**
CLASS	**Insecta**
ORDER	**Lepidoptera**
FAMILY	**Uraniidae**

GENUS **Several, including *Alcides*, *Chrysiridia*, *Uranus***

SPECIES **About 100, including zodiac moth, *Alcides zodiaca*; Madagascan sunset moth, *Chrysiridia riphearia*; Sloane's urania, *Uranus sloanus***

LENGTH
Wingspan: 2–5 in. (5–13 cm), depending on species

DISTINCTIVE FEATURES
Large, colorful, day-flying moths, with wing tails; also duller, pale, night-flying, without tails on wings

DIET
Often plants from family Euphorbiaceae

BREEDING
Typically lepidopteran; eggs deposited on and around suitable host plants

LIFE SPAN
Not known

HABITAT
Largely tropical

DISTRIBUTION
Species-dependent. Madagascan sunset moth: Madagascar; Sloane's urania: South America; zodiac moth: Australasia.

STATUS
Not known

Diverse caterpillars

The host plants of urania moths vary from species to species, but tend to belong to the family Euphorbiaceae. Caterpillars of the zodiac moth feed mostly on the foliage of *Omphalea*, a genus of this family, as do those of the Madagascan sunset moth and Sloane's urania (*U. sloanus*). *Omphalea* foliage is poisonous and when consumed by the larvae may protect them from predation. Urania moth caterpillars are very diverse in shape and appearance. That of the zodiac moth is black with white bands and has a red thorax and legs and yellow prolegs (fleshy legs occurring on the abdominal sections of some insect larvae, but not in the adult). The sunset moth caterpillar has black spatulate spines, whereas the larva of the northern Indian species *Epicopeia polydora* is covered with white cottony filaments. When the caterpillars pupate, they spin a loosely woven silken cocoon.

Structural coloration

A number of urania moths have iridescent coloration, which means that their color changes with the angle of the light. As mentioned earlier, the sunset moth's color varies from green to blue and gold. Usually colors are produced by pigments that absorb certain wavelengths of light and reflect others. Iridescent, or metallic, colors, however, are caused by minute structures in an insect's scales that refract and reflect different wavelengths in different directions. Iridescent colors are consequently known as structural colors. In the wings of butterflies and moths two types of structural coloration are recognized, the morpho type and the urania type. The difference between the two types lies in the construction of the insects' scales, but the effects are the same. The structures in the scales reflect and interfere with light to produce a range of colors, or iridescence. Although the morpho type is named after a genus of butterflies (*Morpho* of tropical America) and the urania type is named after a group of moths, the two types are not confined to either group of lepidoptera. For instance, the glorious birdwing swallowtail butterflies (of the genus *Ornithoptera*, for example) have the urania type of coloration.

VAMPIRE BAT

T HE REAL-LIFE VAMPIRE BAT is quite unlike the vampire of fiction, apart from the fact that it feeds on blood. True vampires, which are native to tropical and subtropical America, feed only on the fresh blood of mammals and birds. Unlike the human-sized vampires of fable, vampire bats are only 14–18 inches (37–45 cm) long, the weight of an adult varying, according to the different species, from ½–2 ounces (15–50 g). Vampire bats' fur is colored in various shades of brown and the animals have no tail. The ears are small, and the muzzle is short and conical, without a true nose-leaf. There are naked pads on the snout with U-shaped grooves at the tip. Thermosensory organs in these grooves enable vampire bats to sense heat that radiates from their prey. The upper incisor teeth are large and razor-edged, well adapted for gently opening a small wound to take blood. The grooved, muscular tongue fits over a V-shaped notch in the lower lip, forming a tube through which the bat sucks the blood of its victim. The stomach is also adapted for liquid feeding, the forward end being drawn out into a long tube. The saliva contains an anticoagulant called draculin that prevents the blood from clotting, enabling the bat to obtain a full meal.

Three vampires

There are three genera of vampire bats, each with a single species. The common vampire bat, *Desmodus rotundus*, the most numerous and widespread of the three, is distinguished by its pointed ears, its long thumb with a basal pad and its naked interfemoral membrane. It has only 20 teeth. The species ranges from northern Mexico south to central Chile, central Argentina and Uruguay and is now one of the most common and widespread mammals in eastern Mexico.

The second species, the white-winged vampire, *Diaemus youngi*, is much less numerous. The edges of its wings and part of the wing membrane are white. It has a peculiar short thumb about one-eighth as long as the third finger and

Vampire bat roosts comprise both males and females. Within the roost, the males defend territories against each other by chasing, pushing, fighting and biting.

has a single pad underneath. The white-winged vampire is the only bat known to have 22 permanent teeth. It is mainly confined to the tropical regions of South America from Venezuela and the Guianas south to Peru and Brazil, but it also has been found on Trinidad and in Mexico.

The hairy-legged vampire, *Diphylla ecaudata*, smaller than the common species, is not well known. It has shorter, rounded ears, a short thumb without a basal pad and softer fur. Its interfemoral membrane is well furred. It has 26 teeth and is unique among bats in having a fan-shaped, seven-lobed outer lower incisor tooth that resembles the lower incisor in the order Dermoptera, the gliding lemurs. This species is found in eastern and southern Mexico, Central America and southward to Brazil.

Communal living

During the day vampire bats roost in caves, old mines, hollow trees, crevices in rocks and old buildings. Colonies of the common vampire may consist of as many as 2,000 individuals, but the average is about 100. The sexes roost together and they may share the caves with other species of bats. They are very agile and can walk rapidly on their feet and thumbs, either on the ground or up the vertical sides of caves. Shortly after dark the bats leave their roosts with a slow, noiseless

The common vampire bat uses its long, well-developed thumb and strong hind legs to bound nimbly over the ground when approaching its prey.

VAMPIRE BAT

CLASS **Mammalia**

ORDER **Chiroptera**

FAMILY **Phyllostomidae**

GENUS AND SPECIES **Common vampire bat, *Desmodus rotundus*; white-winged vampire bat, *Diaemus youngi*; hairy-legged vampire bat, *Diphylla ecaudata***

WEIGHT
½–2 oz. (15–50 g)

LENGTH
Wingspan: 14–18 in. (37–45 cm)

DISTINCTIVE FEATURES
Small ears; chisel-like incisors; well-developed hind legs; heat-sensitive pits in face

DIET
Blood of mammals (*Desmodus rotundus*) or birds (*Diaemus youngi, Diphylla ecaudata*)

BREEDING
(*Desmodus rotundus*) Breeding season: year-round; number of young: usually 1; gestation period: about 210 days; breeding interval: usually 1 year

LIFE SPAN
At least 20 years

HABITAT
Arid and humid regions of Tropics and sub-tropics. Roosts in caves, buildings and trees.

DISTRIBUTION
Mexico south to central South America

STATUS
Common

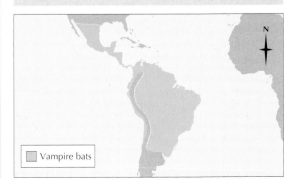

Vampire bats

flight, usually only 3 feet (90 cm) above the ground. The bats attack their victims while they sleep, making a quick, shallow bite with their sharp teeth in a place where there is no hair or feathers. They cut away only a very small piece of skin, making a shallow wound from which they lap the blood without a sound, so that the victim does not wake. Unlike other bats, vampires do not cling with their claws but rest lightly on their thumbs and small footpads, so lightly that even a human is unlikely to be wakened by their actions. The common vampire bat in particular can consume such large quantities of blood that it is barely able to fly for some time afterward.

The common vampire attacks only large mammals, such as horses, cattle and, occasionally, humans. Cattle generally are bitten on the neck or leg and humans are often bitten on the big toe. The white-winged vampire attacks mainly birds, biting the neck or ankle, and occasionally mammals. The hairy-legged vampire appears to prey mainly on birds such as chickens, but it is possible it may also attack some mammals.

In captivity, vampire bats have been kept alive on defibrinated blood, which has had its clotting agents removed to prevent coagulation. One survived for 13 years in a laboratory in Panama.

Finding their prey

Vampire bats detect their prey primarily by using their thermosensory ability and through smell, although, as in other bats, they may use echolocation (locate objects according to the way sound waves are reflected back from them). Because their source of food is large and relatively stationary, vampires do not have the same difficulty in finding their prey as bats that feed on fast-moving insects or even those that catch fish. Like the New World fruit-eating bats, which also feed on stationary food, their echolocation is by pulses having only one-thousandth of the sound energy of those used by bats feeding on insects or fish. It is notable that vampires very seldom attack dogs, presumably because they have more sensitive hearing than larger mammals such as cattle and are thus able to detect the bat's higher sound frequencies.

Year-round breeding

The breeding habits of the white-winged and hairy-legged vampire bats are still a subject of scientific debate. The common vampire gives birth to a single young, occasionally twins, after a gestation period of about 210 days. They breed throughout the year and it is possible there is more than one birth a year. The young are not carried about by the mother, as is the case in most other bats, but are left in the roost while she goes out foraging.

Disease transmitters

The real danger of vampire bats lies not so much in their feeding on the blood of domestic animals and humans, but in the transmission of disease that results from the bites and risk of secondary infections. Vampires can transmit rabies, which may be fatal to both cattle and humans. They also may transmit the disease to other species of bats and may die of it themselves. In Mexico alone it is necessary to inoculate thousands of head of cattle against the disease each year. The disease is always fatal to uninoculated cattle.

Various control methods have been tried in the past to control vampire bat numbers. These include dynamiting the caves where the bats roost and the use of flamethrowers and poison gas. However, such approaches have been found to be largely ineffective in reducing vampire bat populations and are also highly destructive to other, harmless species of bats. Moreover, although such drastic methods may be successful in killing the bats, they do not help to reduce the overall vampire bat population to a level satisfactory to farmers. The only solution to the problem seems to lie in biological control, including sterilization, habitat management and the use of selective chemical attractants and repellents. A research center has been set up in Mexico City for the ecological study of vampire bats and for research into biological methods of control.

The common vampire bat feeds almost exclusively on the blood from livestock such as horses, cattle, goats, pigs and, as in the picture above, donkeys.

VANESSA BUTTERFLIES

A red admiral on brambles. This is a common butterfly in North America, the adult often feeding on the lilac or butterfly bush (buddleia**) or taking sap from trees.**

THE VANESSAS INCLUDE SOME of the most colorful butterfly species in the Northern Hemisphere, a number of them having a worldwide distribution. They are members of the family Nymphalidae. In all species the front pair of legs is reduced in size, the rear two pairs alone being used for walking. They are fairly large butterflies, with a powerful flight. Most of the species resident in North America and northern Eurasia pass the winter hibernating as butterflies, whereas others are continuously brooded in the subtropics and migrate north in summer.

Vanessa caterpillars have branched spines to deter birds and other predators. The pupae, or chrysalises, which are suspended by the tail, are ornamented with gleaming, metallic spots. It was the ornamentation of these pupae that led early collectors to call themselves aurelians, from the Latin *aureus*, meaning golden. Beyond these common attributes, the distribution and life history of the vanessas are diverse, and some of the most familiar species are best described individually.

Red admiral

The red admiral, *Vanessa atalanta*, is found across southern Canada and south through the United States to Mexico and some Caribbean islands. It is also found across temperate Eurasia and has

been introduced into Hawaii and New Zealand. Admirals are represented worldwide, by such species as the Indian red admiral, *V. indica*, and the blue admiral, *V. canace*, which is confined to southeastern Asia. This butterfly is seen on buddleia during warm spring days. The caterpillar eats hops and nettles, creating a hiding-place on the latter by furling a leaf around itself.

White admiral and viceroy

The white admiral, *Limenitis arthemis arthemis*, is found in eastern parts of North America. Its blue-black wings, marked with white streaks, span almost 3½ inches (8.8 cm). Its relative the viceroy, *L. archippus*, displays a form of mimicry in that its dark-veined, orange wings resemble those of the monarch, *Danaus plexippus*. The monarch is foul-tasting to birds, which thus avoid preying on it or on its innocuous looka-like. The viceroy is found along the length of North America east of the Rockies and in Mexico. In southern parts of its range the viceroy is more brown in color and looks like the queen butterfly, *D. gilippus*, which also tastes foul to birds.

Mourning cloak

The mourning cloak, *Nymphalis antiopa*, is found throughout the Northern Hemisphere, as well as on the Andes Mountains. It is called the Camberwell beauty in Britain, where, in spite of its wide distribution, it is a great rarity. It appears that a severe winter is required to induce full hibernation. It has been suggested that the few individuals seen in Britain are not true migrants but stowaways on Scandinavian ships.

Not all North American mourning cloaks hibernate; some migrate south in the fall, returning in the spring to mate. The caterpillars live communally on the food plants, which include willow, poplar, birch, elm and hack-berry. They emerge from the pupae as adults in June or July.

European peacock

This striking butterfly, *Inachis io*, ranges from Britain eastward to Japan. It spends the winter hibernating in dark, sheltered places, such as attics and outhouses, and the caterpillars eat nettles. The peacock has one or two generations a year, depending on the weather. The adult is unusually long-lived: one captive individual lived for 11 months.

Map butterfly

This little vanessid, *Araschnia levana*, is widespread in France and western Europe, although attempts to introduce it into Britain have failed. It is remarkable in appearing in two distinct seasonal forms. Unlike most vanessas, it overwinters as a pupa, and the butterflies that hatch in May are checkered tawny and black and look rather like fritillaries. The larvae from the eggs of these spring butterflies feed and grow rapidly, pupating to produce in July a generation of black-and-white butterflies totally unlike their parents. The length of day during the larval stage

determines which form the mature butterfly will assume. By exposing caterpillars to long or short "days" under artificial light, successive generations of either form can be bred. The larval food plant is entirely nettles.

Question-mark butterfly

Polygonia interrogationis has warm-colored wings of ocherous orange, purplish gray and black. In North America, it is found from Labrador south across the eastern two-thirds of the United States and into Mexico. Elm and nettles are the favored food plants of the caterpillars, which have some of the most spectacular spines of any vanessa. Creamy yellow in color, these develop with each newly molted instar (growth stage), until every long, fine spine bristles with tiny branches.

A slightly smaller, browner relative of the question-mark is the satyr comma, *P. satyrus*, which occurs over the western United States and Canada. Native also to Eurasia, this comma had an interesting history in Britain during the 20th century. Until about 1920 it was confined to a small area in South Wales, but about that time it began to spread over southern and central England. One explanation given is the increased planting of one of its food plants, the hop, which is grown as a flavoring for beer. Other food plants are nettles and elm.

The comma hibernates as a butterfly but remains in the open in woods and hedges, sheltering under leaves instead of seeking shelter in natural hollows or buildings. When its wings are closed, the coloration and irregular outline give the butterfly the appearance of a withered leaf.

VANESSA BUTTERFLIES	
PHYLUM	**Arthropoda**
CLASS	**Insecta**
ORDER	**Lepidoptera**
FAMILY	**Nymphalidae**
GENUS AND SPECIES	**Red admiral, *Vanessa atalanta*; white admiral, *Limenitis arthemis arthemis*; viceroy, *L. archippus*; mourning cloak, *Nymphalis antiopa*; map butterfly, *Anaschis levana*; question-mark butterfly, *Polygonia interrogationis*; others**

LENGTH
Wingspan: typically 1⅕–3 in. (3–8 cm) depending on species

DISTINCTIVE FEATURES
Four walking legs, with front pair reduced in size; often brightly colored upperwings and cryptically colored underwings

DIET
Adult: nectar, sweat, droppings, carrion. Larvae: varied, including nettles, elm.

BREEDING
Eggs laid in batches or singly on appropriate host plant; some larvae communal

LIFE SPAN
May survive several months

HABITAT
Various open or wooded habitats

DISTRIBUTION
Worldwide

STATUS
Mostly common

Adult mourning cloaks emerge in midsummer, but after a brief feed they then enter a period of estivation until the fall, when they feed again before the long winter sleep.

The bold colors of a small tortoiseshell against stonecrop (Sedum). This is one of the most common nymphalids found in western Europe.

Without this cryptic coloration, the butterfly would never escape hungry birds in winter. There are two generations in the summer.

Other North American commas include the eastern comma, *P. comma* (east from the Rockies); green comma, *P. faunus* (northern North America south of the tundra, as well as the Appalachians); hoary comma, *P. gracilis* (central Alaska to northern New Mexico); oreas comma, *P. oreas* (western North America from British Columbia to central California); and gray comma, *P. progne* (northern North America).

Small tortoiseshell
The attractive little *Aglais urticae* ranges across the Eurasian continent to Japan. It is very common in western Europe and can be seen in gardens through most of the spring and summer, because it goes through two generations a year. The butterflies of the second generation hibernate and reappear in spring. Its caterpillars eat nettles.

Thistle butterfly
Named for its caterpillars' food plant, the thistle, *V. cardui*, is the only butterfly species with a worldwide distribution, except Antarctica and Australia. This accounts for another of its common names, the cosmopolitan. In northern Europe and in Britain, where it is called the painted lady, it is a summer migrant, like the red admiral.

The main breeding ground of painted ladies is North Africa, where the emergence of thousands of adults from pupae among the sand dunes marks the spectacular debut of their massed flight toward the Mediterranean. Closer to home, thistle butterflies hatch out in the deserts of northern Mexico and migrate, sometimes in vast numbers, north as far as the Canadian subarctic.

Tasting with their toes
As already mentioned, all the nymphalid butterflies (vanessas, fritillaries, emperors and others) use only the hind two pairs of legs for walking. The stunted forelegs of the males have only two terminal joints and are brushlike. In the females, these legs are more slender with four terminal joints and are only sparsely haired. They are used as sense organs, the terminal joints serving as organs of taste. A red admiral can distinguish, by touching with its forefeet, between pure water and a sugar solution one two-hundredth of the minimum strength detectable by a human tongue. Sweet fluids, such as nectar, are rich sources of energy for adult butterflies in search of mates.

The butterfly diet is, however, not always so fragrant. Both the white admiral and viceroy breed as far north as northern Canada, where nectar sources may not always be sufficient; they make do with feeding on animal dung, rotting fungi and carrion, such as road kills.

VANGA

THE 15 VANGAS or vanga-shrikes form a family of birds confined to Madagascar. They are shrikelike in appearance but, because of the isolation of Madagascar, they have evolved separately, so it is difficult to trace their relationships with other birds. The vangas share characteristics with the wood hoopoes, wood swallows and the shrikes, but as the least-specialized vangas, such as Chabert's vanga, *Leptopterus chabert*, have a similar appearance to the bush-shrikes of the African mainland, it seems most likely that the vangas are derived from bush-shrikes that crossed to Madagascar.

Vangas are 5–12 inches (12.5–30 cm) long, usually black above and white below, but some are brighter. The blue vanga, *L. madagascarinus*, is bright blue on the head, back, wings and tail, with some black on the flight feathers and white beneath. The rufous vanga, *Schetba rufa*, has a black head, white underparts and a rufous brown back. The red-tailed vanga, *Calicalicus madagascariensis*, looks like a finch and is mainly grayish brown above and white below. The face is black with white eye rings, and the male has a black crescent on the breast. In some species the sexes are similar, but the male Bernier's vanga, *Oriolia bernieri*, is black, whereas the female is mainly rufous.

As with other passerine families, such as the Hawaiian honeycreepers and Darwin's finches (both discussed elsewhere), which have evolved in the isolation of islands, the greatest variation of form lies in the bill. Some vangas have relatively simple finchlike or shrikelike bills; others have strangely shaped bills, like the helmet vanga or helmetbird, *Euryceros prevostii*, which resembles a miniature toucan with its exaggerated blue bill that is deeper than the skull, and the sickle-billed vanga, *Falculea palliata*, once thought to be a starling, with a long curved bill. One species fills the ecological niche of a nuthatch, and resembles the latter to the extent that it is called coral-billed nuthatch, *Hypositta corallirostris*. It even looks like a nuthatch. The helmet vanga has a large, casqued bill, recalling those of the hornbills. The sickle-billed vanga has a very long, thin, curved bill. The red-shouldered vanga, *Calicalicus rufocarpalis*, was first described in 1947 but not rediscovered until an expedition to southwest Madagascar in July 1997, when nine singing males were found.

Restricted range

There are three families of birds found only in Madagascar, the asities or false sunbirds, the mesites (discussed elsewhere) and the vangas, constituting probably the least known of all bird families. There are relatively few published records of the habits of the vangas. The endangered Van Dam's vanga, *Xenopirostris damii*, is a bird of primary deciduous forest that lives in northwest Madagascar. In the last 100 years it has been seen at only two sites. Pollen's vanga, *X. polleni*, and Bernier's vanga are also listed as vulnerable. Pollen's vanga has been recorded throughout the rain forest belt of eastern Madagascar. Bernier's vanga is thought to be restricted to undisturbed tracts of the northern rain forest belt. Described as the rarest vanga in the east of the eastern forest, this bird's restricted range makes it particularly vulnerable.

Vangas live in trees, mostly in the fast-disappearing forest, but also in scrubland, mangrove swamps and savanna with scattered trees. Most forage together in small parties of about a dozen, sometimes several species flocking together, but the hook-billed vanga, *Vanga curvirostris*, is more solitary. They usually are seen in the treetops, where they travel through the foliage with great agility, calling to each other with a variety of whistles and chattering calls.

The sickle-billed vanga sometimes utters a range of groans, cries and laughlike sounds. Its local name is voronzaza, or "bird baby," because of its childlike calls.

Acrobatic feeders

Ornithologists believe vangas feed mainly on insects and other small animals up to the size of chameleons. Most vangas feed in the foliage or along twigs and branches, picking off small and medium-sized insects or probing for them in crevices. When foraging, some vangas behave remarkably like tits, agilely flitting among the foliage and hanging upside down to reach awkward places. Some larger vangas hunt like shrikes, waiting on a perch and then dropping on their prey. Chabert's vanga is an aerial insectivore. The white-headed vanga and the hook-billed vanga, among others, include a large amount of vertebrates in their diets, such as tree frogs, chameleons and lizards.

Nesting habits unknown

Most species build open, cuplike nests, often in the fork of a forest tree. There is still much investigative work to be done on the nesting habits of vangas, and even the nests of many species have yet to be found. The sickle-billed vanga, however, has been watched while building its cup-shaped nest of dead twigs about 30 feet (9 m) up in a tree. The female collected and carried the twigs while the male accompanied her. The clutch consists of 3 or 4 white or green spotted eggs.

Vangas and adaptive radiation

As in Darwin's finches (discussed elsewhere), it seems the vangas provide an example of adaptive radiation. It is probable that an ancestor of the vangas arrived in Madagascar, where lack of competition enabled it to evolve into many species, each occupying a certain habitat or having a particular feeding habit. The extent to which evolutionary radiation has proceeded among the vangas is marked: the 15 species are divided into 12 genera. The vangas also have evolved differently shaped bills, but with the limited knowledge of the family it has proved impossible for ornithologists to relate the various vangas to different habitats or feeding habits.

Chabert's vanga is restricted to the island of Madagascar. It frequently forms small flocks and flits between trees, searching for insects.

RED-SHOULDERED VANGA

CLASS	**Aves**
ORDER	**Passeriformes**
FAMILY	**Vangidae**
GENUS AND SPECIES	***Calicalicus rufocarpalis***

WEIGHT
Not known

LENGTH
Not known; other species 5–12 in. (12.5–30 cm)

DISTINCTIVE FEATURES
Male: olive-gray upperparts with rufous shoulder patches; black bib; narrow white band over forehead to behind eye; whitish cheek patch; yellow iris; longish tail rufous below, grayer above

DIET
Small insects

BREEDING
Not known

LIFE SPAN
Not known

HABITAT
Dry scrub with 6½–10 ft. (2–3 m) tall Euphorbia bushes

DISTRIBUTION
Southwest Madagascar

STATUS
Estimated at 30 to 100 pairs

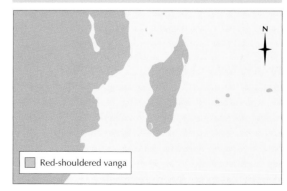

Red-shouldered vanga

VELVET ANT

DESPITE THEIR name, velvet ants are actually solitary wasps. Their common name derives from the marked difference between the male and female of the species. Although the male has wings, the female is wingless and is therefore forced to move about on the ground; in appearance, she closely resembles a large, hairy ant.

The bodies of velvet ants are covered with a pile of short, velvetlike bristles, often patterned in black, bright orange and scarlet. Even the antennae are covered with these short hairs. The 3,000 species of velvet ants are all much alike in color, and most of them live in the hotter, drier parts of the world, especially in the United States. A few species of velvet ants live in temperate latitudes, including Europe, and two species live in Britain.

All velvet ants are parasitic on the larvae and pupae of other insects, including bees and wasps. There are even velvet ants that parasitize other solitary wasps, known as hunting wasps, that themselves prey on other insects. Some of the desert species of velvet ants have a thick covering of long, whitish hairs.

The smallest velvet ant, *Mutilla lilliputiana*, is 3 millimeters long and the largest, *Dasymutilla occidentalis*, of the southeastern United States, is 1 inch (2.5 cm) long or longer and is known as the cow-killer or mule-killer. The so-called large velvet ant of Britain, *Mutilla europaea*, is only about ½ inch (1.3 cm) long.

In many species the male is nearly twice the size of the female, but in the three for which sizes are given here the females are only slightly smaller than the males.

Well-armed and well-defended

Velvet ants are equipped for entering the nests of bees and other wasps; they are not only armed but also armored. They have unusually thick and hard outer coverings. Indeed, many entomologists (zoologists that specialize in insects) have reported experiencing difficulty in pushing a steel pin through the thorax of a dead velvet ant when adding a specimen to their cabinet.

The males are without a sting and can be handled safely. The female, however, has a long and formidable sting, although the power of its venom has often been exaggerated. There is no reason to suppose, for example, that the sting could cause the death of a large quadruped such as a mule or cow. However, the female can inflict a painful sting if handled carelessly, and it is presumably lethal to the insects, such as bumblebees, whose nests the velvet ants parasitize.

Male and female large velvet ants are able to make a squeaking sound by stridulation (rubbing specially adapted body parts together), using a file-and-scraper type of organ. This organ is situated about halfway along the upper surface of the body.

Female velvet ants are usually encountered running actively on the ground near the nests of the species that they victimize. Alternatively, they may be found inside if the nests are opened up. The male velvet ants visit flowers and generally are noticed only by entomologists with specialized knowledge.

Aggressive courtship

In temperate regions male and female velvet ants appear and mate in spring. The naturalist Professor H. M. Lefroy, in describing the mating of tropical species, spoke of the males as powerful insects. Lefroy recorded that when a male finds a female, he "seizes the female by the thorax and flies off; on some convenient spot he mates with her, clasping her firmly to him by his

Despite its antlike appearance, the female velvet ant has mouthparts more typical of a wasp. Pictured is a female velvet ant, D. occidentalis.

Velvet ant larvae undergo complete metamorphosis inside bee or wasp pupae in the host nest, eventually emerging as adult wasps.

VELVET ANT

PHYLUM	**Arthropoda**
CLASS	**Insecta**
ORDER	**Hymenoptera**
FAMILY	**Mutillidae**
GENUS	**Many**
SPECIES	**3,000**

LENGTH
⅛–1 in. (3–25 mm)

DISTINCTIVE FEATURES
Body covered in bristles; color often black, bright orange and scarlet; wasplike mouthparts. Female: antlike, very hard body; wingless; powerful sting. Male: wings.

DIET
Larvae: grubs and pupae of bees and wasps

BREEDING
Details not known; undergo complete metamorphosis in bee or wasp nest; larval stage within host; eggs laid in pupal cells; larvae feed on pupae; emerge as adults

LIFE SPAN
Typically a few weeks or months

HABITAT
Host nests in desert or semidesert

DISTRIBUTION
Mostly in Tropics; a few species in temperate regions

STATUS
Common

forelegs and standing erect on the others … in the frequent intervals the male shook the female with a twisting motion as we would shake a bottle whose contents we desired to mix."

Parasitizing nests

After mating, the female embarks on a search for the established nest of another insect, such as a bumblebee, probably covering considerable distances in the process. She then enters the nest, and is well equipped to resist any attempts to evict her. The female remains in the nest, feeding on the bees' store of honey, and eventually she lays her eggs in the pupal cells or cocoons of the bees, one egg in each cell. The larvae from these eggs feed on the host pupae; then they themselves pupate, emerging later as adult wasps. In temperate latitudes they pass the winter as pupae in the bees' nest, which is abandoned by the bees at the end of the summer.

There is a record of a bumblebees' nest dug out of the ground containing 76 velvet ants and only two bumblebees. This is probably an abnormally high number of velvet ants but it illustrates how effective the female velvet ant must be in coping with the efforts of the female bumblebee, the rightful owner of the nest, to drive her out.

A coat to keep cool

The behavior of both sexes of velvet ants has made it difficult for scientists to study the species. As a result, not a great deal is known about them, and no entomologist seems to have carried out precise experiments to test why velvet ants should have such hairy bodies. Scientists can only assume that the hairs serve as an insulating layer, the evidence for this suggestion being that these insects are essentially desert and semidesert dwellers. By way of comparison, the camel's coat keeps in the heat during the cold nights in the desert and it also keeps out the heat of the sun by day. The female velvet ant, being wingless and forced to run over the hot sand during the day, probably needs the insulating layer against the heat from the ground as well as from the sun. Her mate, in the case of most species of velvet ants, can keep cool by flying or by perching in cooler air. However there are some species in which even the males have only degenerate knoblike wings or are themselves completely wingless.

VENOMOUS ANIMALS

ONE OF THE MOST VALUABLE and widespread abilities in the animal kingdom is that of creating biological toxins. These have evolved in vertebrates and invertebrates, providing even the smallest creatures with potentially lethal weaponry. Specific venomous and poisonous animals are discussed in separate articles.

A distinction can be made between a toxin-producing animal that is merely poisonous, and one that is venomous. The former secretes toxins in its body, which often displays bold warning colors to potential predators, whereas a venomous creature possesses a specialized body part, such as fangs or a stinger, to deliver its toxin into the victim. Spiders and snakes, for example, have hollow or grooved fangs, and a wasp stings with a modified ovipositor (egg-laying organ). Whereas a toxic animal usually uses its toxins in defense, a venomous species may act in defense and offense. Thus, a spider envenomates its prey to subdue its struggles and liquefy its innards, ready for eating; a cobra may deliver a lethal bite into a buffalo merely to defend itself from being trampled. Venom need not be lethal: a female scarab-hunter wasp stings a beetle grub to paralyze, rather than kill it, thus obtaining a live host in which to lay eggs and, later, on which to feed its larvae.

Killer cocktails

In general, defensive toxins contain a narrow spectrum of active ingredients, whereas those manufactured by venomous animals are usually a more complex recipe. The active agents

The venom of Australia's inland taipan, Oxyuranus scutellatus, *contains a blood-clotting agent and is lethal to humans.*

in venom tend to be enzymes: typically these are proteins that act as biological catalysts for cellular metabolism. There are around 20 enzymes that, in various combinations, feature in most of the venoms in the animal kingdom. On average the toxins of venomous creatures are made from a mixture of between 6 and 12 enzymes, the effects of which fall into broad groups. Neurotoxins, typically proteins with a low molecular weight, cause paralysis or affect the nervous system. Hemotoxins and hemolysins affect the blood, often interfering with red cells or clotting agents; hemorrhagins damage blood vessels and cause internal bleeding; myotoxins attack muscle, nephrotoxins the kidneys, and cardiotoxins the heart. Necrotoxins and cytotoxins, typically proteins with a high molecular weight, attack cells and cause necrosis.

In snakes there is a general split between the mainly hemotoxic vipers and the mainly neurotoxic mambas and cobras. The yellow-jawed lancehead, *Bothrops asper*, is a viper. It has large fangs that fold back when the mouth is closed. The lancehead makes a lightning-fast strike that constitutes more of a stab than a bite; having injected venom, it releases the victim instantly. One reason for this is that the hemotoxins and necrotoxins in the venom will cause the victim immense pain, and by keeping clear of the victim until its violent death throes have ceased, the snake avoids injury to itself.

Venom delivery

Methods of venom delivery are almost as diverse as the venomous creatures themselves, ranging from fangs (snakes, spiders), stingers (scorpions, bees and wasps), spines (fish) and pincerlike mouthparts (centipedes and ants), to squirting devices (millipedes and ants) or venomous spurs (duck-billed platypus). All these methods share basic similarities. There is usually a gland or glands that produces the venom, a delivery system and a duct connecting the two. Spitting cobras (genus *Naja*) have small fangs fixed at the front of their jaws, with the discharge orifice facing outward, allowing their owner to fire streams of venom directly into the eyes of a potentially dangerous intruder. Prey is rarely treated in this manner, though, as most cobras have fast-acting, neurotoxic venom and hold onto prey after they have struck until the toxins have started to take effect. The Gila monster, *Heloderma suspectum*, one of only two venomous lizards, delivers its venom via grooves in its teeth. To improve the effect of its bite the lizard must chew the toxin into the victim.

The carnivorous cone shells (genus *Conus*) of the Indian Ocean and South Pacific move much more slowly than the worms, mollusks and fish on which they prey, so these gastropods use a harpoon laden with a paralyzing venom. Held in the tip of the proboscis, this hollow, barbed tooth is shot into prey while the fast-acting neurotoxin is squeezed through the tooth cavity. Each barb is used only once and is replaced with the next in a series of teeth held in reserve. Once it has successfully envenomated its prey, the cone shell swiftly extends its proboscis to envelop and consume its victim. Unwary bathers and divers that step on the shell discover its potency to their cost: the venom can stop the human heart in a mere 10 minutes. Today this potentially lethal neurotoxin is being researched for its efficacy, in much reduced doses, on brain-damaged stroke victims.

The seabed is also home to scorpion fish (genus *Scorpaena*), stonefish (genus *Synanceia*) and weever fish (genus *Trachinus*). These lie in the sand, exposing only their camouflaged upperparts. Spines in their dorsal fins are modified to inject venom. Pressure, such as from a human foot, causes the tip to snap off. Venom floods from a gland, up the spine and into the victim. Even if humans do not die from the venom, bacteria introduced into the wound cause it to fester.

A jellyfish is not built to struggle with prey, so its venom must very rapidly subdue a victim. Accordingly, it possesses thousands of specialized cells known as nematocysts, which are coiled harpoons laden with venom. Upon contact with ·chemicals in the victim's skin, be it fish or mammal, each nematocyst fires its harpoon. The tentacles stick to the skin, enabling more nematocysts to fire. Any attempt to remove the stingers merely enhances their effect. In some jellyfish species, the mass dosage of neurotoxin is enough to stop the heart of an adult human, let alone a fish prey. The box jellyfish or sea wasp, *Chironex fleckeri*, found in waters off Australia and Indonesia, uses a passive hunting strategy, swimming among

Sea snakes hunt fish. Though adapted for the chase, they are not so rapid as their prey, so they need very powerful venom to prevent the wounded from escaping.

Death by proxy

Many venoms exploit the victim's own pathological response to envenomation. For example, neurotoxins are present in almost all scorpion venoms. When a scorpion stings a human, the neurotoxins in its venom disrupt the normal electrical function of the victim's nerve cells. The victim's body then overreacts, manufacturing too many neurotransmitters and generally destabilizing its own natural equilibrium. All these secondary effects can result in excruciating pain, nausea, fever and, in rare cases, cardiac or respiratory failure. Thus it is strictly not the venom itself, but the added effect of its neurotoxins that is the deadliest agent (although additional ingredients, such as histamine or serotonin, have a direct effect on blood vessels). Sometimes it is not the venom's toxicity that is dangerous in itself, but other substances contained in it. The venom of the white-tailed spider (family Gnaphosidae) is not deadly, but it carries bacteria that can cause tissue damage around the area of the bite.

prey and waiting for them to brush past. Periodic swarms of these jellyfish are a hazard to swimmers. The Portuguese man-of-war (genus *Physalia*) has a specialized, air-filled polyp that floats at the water's surface, suspending beneath it the rest of the colony: digestive polyps and long, venom-laden tentacles that may trail up to 165 feet (50 m). This curtain sweeps through the surface waters of the open ocean to net prey.

Warning colors

Those species that are poisonous, rather than venomous, include the soft-bodied and vulnerable frogs, toads and newts. These secrete toxins in their flesh or skin. Different portions of such animals may vary in toxicity. In some frogs and toads, merely being placed in a predator's mouth causes the poison to froth out in such quantity that the attacker backs off before biting. It does the amphibians no good, however, if the predator gets as far as taking a firm bite before recoiling; self-preservation relies on warning the predator not to attack in the first instance. Accordingly, many toxic amphibians are brightly colored or patterned. Predators learn to ignore the warnings at their peril. When threatened, many newts of the genus *Taricha* adopt a response known as the unken reflex, bending up the head and tail to expose bright warning colors on the belly. The same tactic is employed by the fire-bellied toad (genus *Bombina*).

In nature, bright coloration can be used as a warning or an invitation, but in certain birds of Papua New Guinea it serves both functions. The feathers of the colorful songbirds known

The saliva of the blue-ringed octopus contains maculotoxin, a neurotoxin that paralyzes its crustacean prey.

locally as pitohui contain secretions of batrachotoxins, which irritate the lungs. Any predator that tries to eat a pitohui finds it chili-hot and unpalatable. Equally, the unpleasant smell the birds give off deters predators that hunt by use of smell.

Among the deadliest animal poisons are those used by the spectacular poison-dart frogs (genera *Dendrobates* and *Phyllobates*). These secrete cocktails of toxins including batrachotoxins and several other unique alkaloids (complex nitrogenous compounds derived from plants). The tiny frogs, which live in rain forest trees, flaunt their jewel-like skin colors and patterns, safe in the knowledge that all potential predators will give them a wide berth. (Almost all: *Leimadophis epinephelus*, a snake, seems to be immune to the frogs' toxins and is probably their only significant predator.)

A single poison-dart frog may yield as much as 1,900 micrograms of poison, equivalent to several times the lethal dose in humans. Colombian forest peoples daub the poison on the blowpipe darts they use to hunt monkeys and other game. The buildup of toxins in the bodies of the frogs is believed to result from their largely ant-based diet. The ants' bodies are high in alkaloids, and these are passed on to the frogs. Captive poison-dart frogs that are reared on other foods lose their toxicity. Medical research institutes have investigated the use of dendrobatid batrachotoxins in anesthetics and heart drugs.

Another highly colorful killer is the blue-ringed octopus, *Hapalochlaena maculosa*, of the Indian Ocean and South Pacific. This small species can cause its cobalt-blue rings to glow intensely at will, threatening would-be predators with a very potent venom. In fact, it has two venoms; one is reserved for the crabs and other crustaceans on which it preys; the other is

The dorsal spines of a scorpion fish can push deep into a victim, heightening the efficacy of the venom that they inject.

used strictly for defense. Both are manufactured from bacteria and are inherited: the octopus's offspring are venomous even before they hatch from the egg.

Venom toxicity

The question of which is the most venomous animal species is relatively meaningless because there are so many variables. The varying number of active chemicals in each venom and their relative effect on each animal class (insect, mammal, bird, and so on) mean that any one venom may have dramatically different effects depending on the victim. The venom of the Australian funnelweb spiders (genus *Atrax*), for example, has little effect on most mammals, but it proves deadly to the higher primates, so a dose that spares a mouse can quite easily kill an adult human. Furthermore, the quantity of venom produced varies, as does the amount injected in a delivery. Many animals, such as scorpions, can regulate the volume of venom they release in each attack. The depth and location of the envenomation may also have a bearing on its effect. Luckily, few delivery systems can penetrate beyond the subcutaneous level of human skin, restricting the extent to which venom can circulate throughout the body.

In tests to gauge venom toxicity, two scales are used in combination: the average venom yield and the LD50 and LD100 indices (the amount of venom required to kill 50 and 100 percent of test subjects respectively). Typically, venom potency is expressed in its lethality on laboratory mice.

However, these tests are highly specific, and small changes to the method whereby the venom is collected, stored and reconstituted can affect the results. Most venoms begin to break down soon after extraction, so a test that stores the venom in ideal conditions and injects it intravenously (thus mimicking real life) will give different results than a less scrupulous test.

An animal's behavior affects how dangerous it is. The Brazilian *Phoneutria* spiders are aggressive toward humans and are responsible for up to 60 percent of spider bites reported in some South American hospitals. At the other end of the scale are the sea snakes (family Hydrophiidae), especially those from the genera *Hydrophis* and *Enhydrina*. These snakes of the Indian Ocean and South Pacific have some of the most virulent venoms known to science. They are, however, reluctant to bite in defense, and experienced field researchers may handle them safely. In reality, just as each species' venom is different, so are the circumstances of each envenomation, and the risk that the venomous strike will prove effective.

Neutralizing venom

Treatment for a serious envenomation requires antivenin, a serum containing antibodies. Biologists create antivenins by subjecting an animal, often a horse, to increasingly high doses of venom, and then collecting the antibodies that the horse produces in response. Some antivenins are species-specific, while others are polyvalent. There are still, however, many venomous creatures for which no antivenin exists, and some humans are allergic to treatment. Of the many thousands who die from snakebite every year in Asia alone, most die because they live and work far from the nearest first-aid station.

VESPER BAT

THERE ARE 980 SPECIES OF bats in 17 families. Nearly one-third of the species are in one family, the Vespertilionidae. Specialists usually refer to them as vespertilionids, but generally speaking they are known by the simplified term of vesper bats. Because most of the 980 species of bats are seen in the evening, this name is hardly specific; yet it is a convenient name for the small, commonplace, insectivorous bats seen particularly in temperate latitudes but found all over the world except at the poles.

Vesper bats mainly are small. They have a total length of 1¼–4 inches (3–10 cm), with wingspans of about 5–15 inches (12.5–37.5 cm). Most of them are various shades of brown, sometimes gray or black, and a few are yellow, orange or red. Some, such as the spotted or pinto bat, *Euderma maculatum*, of the United States and Mexico, have white patches. Vesper bats typically lack nose leaves but have an earlet. The ears, as in the European long-eared bat, *Plecotus auritus*, the American long-eared bat, *P. macrotis*, and Townsend's big-eared bat, *Corynorhinus townsendii*, may be up to 1½ inches (3.8 cm), half the head and body length.

Sleeping quarters

Most vesper bats spend the daylight hours in cavities in hollow trees, under loose bark, among foliage, or, more likely, in rock crevices, buildings, tunnels, mine shafts and natural caves. They usually hang by their hind feet and prefer to sleep on a vertical face rather than hanging free by the toes. Some are solitary but most roost in small groups or in colonies. The sexes often roost separately, especially when the females are giving birth or have young. Roosting places often are traditional, with the same colony using a particular place year after year. Some species move from a summer feeding ground to winter quarters, which may be several miles away. The red bat, *Lasiurus borealis*, and hoary bat, *L. cinereus*, make long migrations southward at the end of summer, returning the following spring.

Highly flexible wings

A bat's skeleton is light and the structure and design of its wings make it highly maneuverable in the air. The arm, which carries the wing membrane, consists of a short upper arm and a

Vesper bats usually hibernate in caves or hollow trees, although many choose buildings as a site in which to roost. The bats shown here are pallid bats.

Although most vesper bats produce only one offspring, the red bat, Lasiurus borealis, *averages three. The mother and young above are in their daytime roost.*

long forearm with a single bone, the radius, and a compact wrist with several of the bones fused together. The finger bones are greatly elongated, especially the third and fourth, which support most of the wing membrane. The thumb is short and bears a claw, used in climbing or walking. The wing membrane runs from shoulder to wrist, over the fingers and backward along the side of the body to the ankle. This membrane is made up of a double layer of skin between which are such slender elastic strands and fine muscle fibers that the wing collapses and folds up easily when not in use. While on the wing, however, a bat has full control through the tendons, which are worked by the arm muscles, controlling all the joints; bats are the most maneuverable of all flying animals. Among vesper bats there are the long-eared bats that fly through foliage and hover to pick insects off leaves and, at the other extreme, the slender-winged noctule, *Nyctalus noctula*, that often catches high-flying insects in flight.

While the bat is flying, its wing acts as a ventilator; the network of fine blood vessels allows the blood to cool down so the bat does not become overheated from the exertion of flying. When the bat lands and folds its wings, the flow of blood to the wings is cut down, and as the surface area is reduced through the folding of the wings, the heat is retained in the bat's body.

Feeding on the wing

Nearly all vesper bats are insect eaters. They have sharp teeth, the molars having a W-shaped pattern of cusps for chewing. One species, the

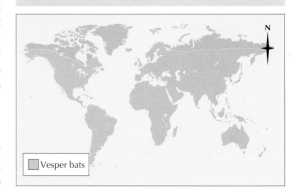

Vesper bats

fishing bat, *Pizonyx vivesi*, of California eats fish and small crustaceans, and a few other species are suspected of catching fish. The desert bat, *Antrozous pallidus*, of North America catches insects near the ground and also captures scorpions and lizards. Most vesper bats eat only insects, which they catch in flight. Small insects are chewed and swallowed straight away. The bats commonly catch larger prey such as moths and beetles, which are briefly pouched in the interfemoral membrane, between the tail and the hind legs, and then rapidly transferred to the mouth. It is indeed remarkable that vesper bats are still able to echolocate (emit sounds and analyze their surroundings according to the way the sound waves are reflected back to them) while holding a large beetle or moth in the mouth.

Vespertilionids come out at variable times in the evening, from just before sundown to almost dark. Some are still seen on the wing just before dawn, but the night is divided into alternating spells of hunting and resting to digest their food. Some species, such as the noctule, have only two periods of hunting, each lasting an hour.

Delayed fertilization

Most of our information about the breeding of vesper bats is from species in temperate regions, where the bats hibernate. In these areas, mating usually takes place between August and October, the sperms being stored in the female, and it also may occur again in spring. All fertilization is in spring, and after a gestation of 40–70 days, or even 100, according to the species, the babies are born from late May to July. In the Tropics, fertilization follows immediately after mating. The number of babies at a birth usually is one or two, but there may be four.

Few predators

Because they fly by night, bats probably have few predators. They are most vulnerable to predators when they emerge from their roosts at twilight, at which time they can be picked off by raptors, such as hawks. In Africa, the bat hawk specializes in this technique. With this exception, few predators manage to make bats a large proportion of their diet. In the middle of the night, owls might catch them occasionally, but by far the greatest threats to vesper bats in most countries are humans and domesticated cats.

Long-lived species

As a result of banding hibernating whiskered bats, *Myotis mystacinus*, in some Dutch caves, scientists discovered that 40 percent die in the first 6 months of life and beyond this the average expectation of life is about 4½ years, although some may live for up to 30 years. This is a very long time compared with other small mammals of similar size, such as shrews and mice, which live to about 5 years at the most. The bat's effective life is, however, very brief. A noctule, for example, spends the winter hibernating. In the remaining months it is on the wing for only about 2 hours in every 24.

The noctule is the most widely distributed vesper bat. Its range includes most of Europe, temperate and Southeast Asia, Algeria and possibly Singapore and Mozambique.

VESPULA WASP

These vespid wasps are roosting by hanging onto to grass stems by their jaws. Of all the inhabitants of a colony, the queen alone will survive the winter.

VESPULA WASPS ARE the insect that most people think of when they hear the term wasp, and they include the yellowjackets that trouble summer picnickers. Strictly speaking, the term applies to the genus *Vespula* in the family Vespidae. These species are all social, sharing the duties in a colony ruled by a single queen. Some nest in trees, others below ground. All have a striped abdomen, the colors varying according to species, and all can deliver a painful sting. The family Vespidae also includes the paper wasps, *Polistes*, and the hornets, *Vespa*. (Confusingly, the name hornet is also used to refer to those *Vespula* species that nest above ground.) The Eurasian hornet, *Vespa crabro*, is a large, striped species with a particularly painful sting. It has been introduced to North America.

The common wasp, *Vespula vulgaris*, and the German wasp, *Vespula germanica*, are equally common in Eurasia, and are so alike that the workers are difficult to distinguish, although the queens can be separated by the pattern of their yellow-and-black markings. Other *Vespula* wasps include the red wasp, *V. rufa*, which nests underground. The tree wasp, *V. sylvestris*, and the Norwegian wasp, *V. norvegica*, hang their nests in trees and bushes. The cuckoo wasp, *V. austriaca*, is found in both North America and Eurasia. In this species the queen enters the nest of a red wasp, kills some of the workers and supplants the queen. The parasitic invader's brood is reared by the red wasp workers, the offspring of the parasite consisting entirely of fertile males and females.

Papier mâché nests

The history of social wasps' nests, such as the subterranean nest of *V. vulgaris*, really begins in the fall of the year before their construction, when the large queens leave the nests where they hatched, mate and then hide themselves, to pass the winter in hollow trees, sheds and attics. The queen finds a rough beam or piece of sacking, clamps her jaws onto a fiber and hangs there unconscious for 6 or 7 months.

The queen emerges in late spring and seeks a crack in the ground or an old mouse's hole running under a tree root. Just below this she digs out a chamber, removing the soil in her jaws. Then she flies repeatedly to and from a fence post or dead tree, each time bringing home a little pellet of paste made by rasping away the wood and moistening the mouthful with saliva. On a still day, the chewing action on a fence post may be audible to humans several feet away. The queen plasters a mouthful of wood pulp onto the underside of the root, where it hardens to form a paperlike substance. She fixes a small curved canopy to this foundation and makes a paper stalk, which descends from the center of the canopy. A cluster of hexagonal cells, also made of paper material, is then built around the stalk, with their open ends downward. The queen lays an egg in each and then encloses this first comb in a bag of the paperlike substance about as large as a golf ball, with a hole at its lowest point.

Growth of the nest

During the nest construction and egg-laying, the queen feeds regularly on nectar. When the eggs hatch into small white larvae, she divides her time between feeding them on the juices of chewed-up insects, for they are growing and so require a protein diet, and adding more cells to the comb, enlarging the enclosing bag as she does so. By the time the larvae from the earliest eggs have passed through the pupal stage to produce the first workers, she may have added a story to her house, built below the first and suspended on small paper stalks. To make room for the growing nest, the queen may have to excavate more soil and carry it away.

VESPULA WASPS

PHYLUM	**Arthropoda**
CLASS	**Insecta**
ORDER	**Hymenoptera**
FAMILY	**Vespidae**
GENUS	***Vespula***

ALTERNATIVE NAMES
Yellowjackets; hornets (erroneously)

LENGTH
Usually ⅖–1 in. (1–2.5 cm)

DISTINCTIVE FEATURES
Large wasps, usually with black-and-yellow striped abdomen; social structure involves queen, worker and male castes

DIET
Sweet foods including nectar, fruit, jelly

BREEDING
Queens and males mate in fall; queen builds nest and lays eggs in spring; larvae become workers, later taking over building and feeding work while queen lays eggs

LIFE SPAN
A few weeks to several months

HABITAT
Nests in trees and buildings as well as underground

DISTRIBUTION
Worldwide except polar regions

STATUS
Common

This close-up of the head of a queen German wasp shows the powerful mandibles, with which she mashes wood into pulp to begin the nest. Also visible are the compound eyes and the sensitive antennae.

When the worker wasps, which are nonreproductive females, appear in quantity, they take over from the queen the job of extending their home. New stories are added, one below the other, increasing to the maximum diameter of the nest and then decreasing again to maintain the roughly spherical shape. Quantities of soil are excavated by the workers and wood pulp is imported for construction. The root anchorage is strengthened as the bulk and weight of the nest increases, and struts and stays are made between it and the surrounding earth.

By now the queen stays at home, fed by her sexless daughters, who also must bring home animal food for all the growing larvae. As each cell is completed she places an egg in it, until a population of as many as 5,000 wasps, more in very large nests, is built up and maintained.

The total number of the queen's offspring that hatch, live and die in the service of the nest during a summer may be five times that number.

When completed, the nest is a hollow sphere 8–9 inches (20–22.5 cm) wide, containing 6 to 10 horizontal combs that extend more or less right across it. The inside of the nest is continually being nibbled away and repulped. Fresh pulp is also added to the outside and to the expanding combs inside, so the whole structure is constantly changing.

Vespula squamosa, which is found from the eastern United States south into Central America, may make either a subterranean or an aerial nest. The former are comparable in size to those of *V. vulgaris*, being typically 8 inches (20 cm) across and sited 9 inches (22.5 cm) beneath the surface of the ground. It may, however, build much larger nests above ground level. One such construction, possibly the amalgamated work of three independent colonies, measured 8⅓ feet (2.5 m) in height.

Feeding and growing

The workers feed on nectar and fruit juices and also accept drops of liquid exuded by the larvae. The larvae and queen are fed by the workers. Wasps destroy great numbers of bluebottle flies in the process of keeping the larvae fed.

The larva, a white, legless grub, maintains its position in the upside-down cell by pressing its body against the sides. When it is fully grown it closes the cell by spinning a papery cover across the mouth. During the larva's life, its excrement accumulates at the end of the intestine and is

A German wasp grooms its hindmost legs while perched on an apple. Orchards attract Vespula *wasps in droves to feast on the sweet fruit.*

voided all at once in the final larval skin when the larva changes into a soft, white pupa. The wasp emerges 3–4 weeks after the eggs are laid.

Fall breeding season

Late in the summer, a generation of males and functional females is produced. The latter are the queens, similar to workers but larger; the males are about the same size as the workers but have longer antennae. Eggs that yield workers and queens are always fertilized by spermatozoa from the store that the queen acquired at mating and keeps in her body. Males are produced from unfertilized eggs, from which the queen withholds sperm as she lays them. After mating, the males die and the queens hibernate. At summer's end the workers become less active and cease to maintain the economy of the nest, and they and the old queen die with the first frosts.

Guests and parasites

A hoverfly (*Volucella* spp.) enters wasps' nests and lays its eggs without any interference from the wasps. Its curious, prickly larvae play a useful role in the nest as scavengers, living in the midden below the nest, where dirt and dead bodies accumulate, and also entering vacated cells and cleaning out the deposits of excrement. This helps in making the cells available for reuse. The larvae of the moth *Aphomia sociella* also live as scavengers in wasps' nests. Late in the season,

when the nest is running down, they invade the combs and devour the grubs and pupae. The larva of a rare beetle, *Metoecus paradoxus*, lives parasitically on the grubs of wasps in underground nests. It is at first an internal parasite, but later emerges and devours its host. *Metoecus* apparently invades only the nests of the common wasp, never those of the German wasp.

Formidable weapon

The wasp's sting is really an ovipositor, or egg-laying organ, that has become transformed into a tiny hypodermic needle supplied by a venom gland. The eggs are extruded from an opening at the base. Wasps sting if they are squeezed or restrained, as when they accidentally crawl inside someone's clothing. They also attack and sting if the nest is interfered with or even simply approached. The inhabitants of large, well-populated nests are more aggressive than those of small ones. Unlike a bee, which stings only once and then dies, leaving the barbed sting in the victim's flesh, a wasp retains its ovipositor and may sting time and time again.

The main active constituents of wasp venom are histamine and apitoxin. Traditional remedies, such as baking soda and ammonia, were based on the mistaken idea that the venom is an acid of some kind, and they are ineffective. The venom is rarely dangerous unless the victim is hypersensitive to it, in which case it may be lethal.

VINE PEST

WHEN EUROPEANS colonized North America, they started a chain of events that nearly ruined the vineyards of California, Europe, South Africa and Australia. Native to the southeastern United States, the vine pest is a tiny insect related to the aphids, which are also known as plant lice, greenflies and blackflies. The vine pest, *Phylloxera vitifoliae*, is sometimes referred to as the vine mite, vine louse, graperoot louse or grape phylloxera, although it feeds on the sap of a wide range of other plants as well as vines.

The vine pest was unwittingly introduced into Europe between 1858 and 1863, when vine growers were experimenting with vine species imported from North America. By 1885 the pest had reached Algeria, Australia and South Africa. It also reached California about the same time, probably taken there on vines from other parts of the United States east of the Rocky Mountains that had been imported from Europe.

Close call for vineyards

From early times only one species of vine has been used for winemaking in Europe. This is *Vitis vinifera*, a native of the region bordering the Caspian Sea, and it has proved extremely susceptible to attacks of the introduced vine pest. The presence of the aphid on a vine is shown first by the stunting of the plant itself and then by the reduction in the size and number of the leaves. In some cases the leaves become discolored and galls form on their lower surfaces. At the same time, knotlike swellings, known as tuberosities, are found on the smaller roots. These blacken and cause the roots to die and decay. The growth of the grapes is arrested and the fruits wrinkle. The vines are weakened and rendered more susceptible to winter injury. At its peak, the vine pest ruined 2½ million acres (1,012,500 ha) of vineyards in France.

A complicated life history

Except that the vine pest threatened to wipe out viticulture in Europe and elsewhere toward the end of the 19th century, the species is remarkable only for its complicated life history. After mating, the female vine pest aphid lays its egg on the bark of the vine. Each egg passes the winter on the bark and hatches in the spring, producing a wingless female called a stem mother, also known as the fundatrix, or foundress. The stem mother crawls into a leaf bud, where it causes a gall to develop on the young leaf. Inside the gall it lays as many as several hundred eggs, which develop into further wingless females called gallicolae, or gall-dwellers. These multiply during the summer, giving further gall-forming generations that in turn infest other leaves. Later in the season the gallicolae produce wingless females of another kind, the radicicolae, or root-dwellers, which go down to the roots.

After producing several generations of their own kind, the radicicolae give rise to winged females, which fly in late summer to other vines. There they lay two kinds of eggs. Small eggs produce males and larger ones produce females, both sexes again being wingless. The mouthparts and digestive systems of this latest batch are not developed so they do not feed, but they mate, and each female lays a single egg. These are the eggs that overwinter and form a new generation of fundatrix females, starting the whole complex series again. This is the typical, complete life history of a vine pest on its natural and native

There are several generations of wingless females (left, above) before the winged females finally emerge in late summer. Vine pests' mouthparts are adapted for sucking sap and they have the capacity to multiply extremely rapidly.

Spraying vineyards with the chemical Thiodan to control vine pests. Growers are careful not to move machinery between infested and clean areas, since the first-instar nymphs, or crawlers, are capable of hitching a ride.

North American host plants. When transferred to European vines, the radicicolae are the principal form, and they seem to be able to hibernate through the winter and reproduce their own kind indefinitely.

How the pest is controlled

The French vineyards are still productive; they were saved by intensive entomological research. There are many species of vines native to North America, and these have varying degrees of resistance to the vine pest. In some the roots are actually toxic to the pests. When European vines are grafted onto stocks of these resistant North American plants, the radicicolae are unable to thrive on their roots and the *Vitis vinifera* scions escape their ravages. *Vitis riparia*, *V. rupestris* and *V. berlandieri* are three American species of vines that are suitable for grafting, and hybrids between them and *V. vinifera* are also extensively used. As a precaution, growers destroy any wild grapevines that spring up near the borders of commercial cultivars.

In the United States grape-growing industry *V. vinifera* is cultivated on resistant stocks, as in Europe. Some resistant species of vines, producing fruits of a variety of different kinds and flavors, are also grown, including *V. labrusca* and *V. rotundifolia*.

Bugs fight back

Growers in North America still have a fight on their hands, however, despite the successes with grafted vines. Problems arise when a vine root is planted that has weak pest resistance (a congenital defect). It only takes a minuscule number of vine pests to infest the root successfully for them to multiply rapidly and raise an immense new, destructive generation. The grower must then grub up the vineyard and invest in a new, resistant root strain, at prohibitive expense. It is estimated that vine pests caused more than $1 billion worth of damage to California vineyards during the 1990s.

VIPERFISH

ALTHOUGH THEY ARE CONSIDERED harmless to humans, the deep-sea viperfish are fearsome in appearance. They have long, fanglike teeth that are slightly barbed at their tips and project on either side of the jaws. The body of a viperfish measures up to about 14 inches (36 cm) long, but its slender build makes it appear longer, and it is only slightly thicker behind the head than in the tail. The head is small but has a strong lower jaw. The fins, including the tail fin, also are small, the pectorals being smaller than the pelvics. There is a small adipose (fatty) fin just in front of the tail fin, and opposite this on the underside are two small anal fins set close together. The most prominent fin is the first dorsal, just behind the level of the rear end of the pectoral fins, featuring a long, whiplike spine formed from the fin's first ray. The spine is about half the fish's length and carries a light organ at its tip. A double row of light organs runs along both sides of the lower body.

Worldwide in the deep
Viperfish live to depths of about 9,000 feet (2,750 m) and perhaps deeper in the Atlantic, Pacific and Indian Oceans and in the Mediterranean, though they may migrate closer to the surface during the night. There are eight species, all belonging to the genus *Chauliodus*. Three species inhabit the Atlantic Ocean. The dana viperfish, *C. danae*, measuring about 6 inches (15 cm), is found on both sides of the Atlantic as well as in the Caribbean. The slightly smaller *C. minimus* also occurs in the western and eastern Atlantic, while *C. schmidti*, at about 9 inches (23 cm) long, is restricted to the east. Three more species are found in the Pacific Ocean. *C. barbatus* and *C. vasnetzovi* live only in the southeast Pacific, around Chile, whereas the Pacific viperfish, *C. macouni*, occurs in the northwestern and eastern regions of the ocean. One species, *C. pammelas*, is found in the Indian Ocean, around Arabia and the Maldives. The eighth species, Sloane's viperfish (*C. sloani*), is found in the Atlantic, Pacific and Indian Oceans and in the western Mediterranean.

A glimpse of another world
Usually, scientists can study deep-sea animals only when these creatures' dead bodies are brought to the surface in nets. In the case of viperfish, however, at least one scientist has had brief sightings of the fish in its element. American naturalist and explorer William Beebe

Chauliodus barbatus *is limited to the southeast Pacific and occurs at depths of more than 6,500 feet (2,000 m). The huge eyes are an adaptation to the dim light at such depths.*

Food is scarce in the deep, so no meal can be passed up, even if it is bigger than the diner. Consequently, viperfish have evolved a hinged skull and expandable stomach.

VIPERFISH

CLASS	**Osteichthyes**
ORDER	**Stomiiformes**
FAMILY	**Stomiidae**
GENUS	***Chauliodus***
SPECIES	**8, including *C. sloani* (detailed below)**

LENGTH
13¾ in. (35 cm)

DISTINCTIVE FEATURES
Slender, iridescent silver body; strong lower jaw; double row of light organs along either side of ventral (lower) area; light organ at end of whiplike dorsal (upper) spine

DIET
Midwater fish and crustaceans

BREEDING
Egg-laying; eggs pelagic (oceanic)

LIFE SPAN
Not known

HABITAT
Deep oceanic waters to depths of more than 3,280 ft. (1,000 m); may move up to near-surface waters at night

DISTRIBUTION
Warm and temperate regions of all oceans; western Mediterranean; absent from some areas, including parts of southern central Atlantic, northern Indian Ocean, eastern Pacific north of equator

STATUS
Not threatened

saw one through the window of his bathysphere during his descent to a depth of 3,028 feet (923 m) in 1934. In his book *Half Mile Down*, published in the same year, he spoke of seeing a red prawn that was pounced on by a "really fearsome" viperfish that shook it for a moment and then swallowed it. He also spoke of the viperfish's stomach, which can stretch enormously as if it were made of rubber.

The viperfish's body, with the second dorsal fin and the anal fins set far back, recalls the pike (*Esox lucius*), a freshwater fish that lunges swiftly at its prey, seizing it in a wide mouth armed with large teeth. Unlike the viperfish, however, the pike has no light organ at the end of a whiplike dorsal spine. This organ seems to be used as a lure to tempt prey within reach, and observers aboard submersibles have seen viperfish lurking in midwater with their lure held forward above the mouth, waiting for prey to come in range. The viperfish also has other light organs about its head, which also presumably entice prey to within reach of its mouth.

Expandable throat

In the early 1950s, the scientist V. V. Tchernavin showed that the first vertebra behind the viperfish's head is large and has broad surfaces for the attachment of strong muscles. The backbone, immediately behind this, is supple. The heart is situated well forward and lies between the bones of the lower jaw, as do the gills. In swallowing prey, the muscles attached to the first vertebra pull the head up, and, with the mouth opening at the same time, the lower jaw is shot forward, so the head seems almost to part company with the body as the throat opens. The effect is to give a wide and clear passage into the gullet. At the same time the heart is carried forward and the delicate gills outward, so they are not damaged by prey, even large prey, entering the throat. When the prey has been swallowed, the head,

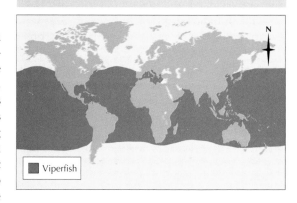

Viperfish

jaws, heart and gills all return to their normal position. It is reasonable to suppose that with so much disturbance of the vital organs, the swallowing action must be rapid. Digestion also seems to be quick because most of these fish have empty stomachs when they are caught.

VIREO

Vireos are small, dull-plumaged song-birds of the New World. The 52 species, together with the little-known shrike-vireos and peppershrikes, make up the family Vireonidae. In the 19th century they were known as greenlets but this name is now used only for 15 species of tropical vireos of the genus *Hylophilus*. Vireos range from 4–7 inches (10–17.5 cm) long and usually are olive green or gray above and whitish or yellowish underneath. The bill is slender and sometimes bears a small hook.

The best-known species is the red-eyed vireo, *Vireo olivaceus*, 5 inches (12.5 cm) long, olive green above and whitish below. The top of the head is slate gray and a white line with a black border runs over the red eye. The red-eyed vireo breeds from central Canada to the shores of the Gulf of Mexico and across much of South America, but is missing from most of the western half of the United States. One of the brightest vireos is the yellow-throated vireo, *V. flavifrons*, of the eastern United States. It is olive green above with two white wing bars, a yellow eye stripe and yellow on the chin, throat and breast. The greenlets are among the smallest vireos. The gray-headed greenlet, *Hylophilus decurtatus*, is 4 inches (10 cm) long with a gray head and a white ring around the eye. The body is olive green above and white and yellow underneath.

The vireos range from central Canada to northern Argentina, including the Caribbean and the Bahamas. The Noronha vireo, *V. gracilirostris*, is confined to the tiny island of Fernando de Noronha, 250 miles (400 km) off the Brazilian coast.

Persistent songbirds

Except for some species, such as the arboreal (tree-living) red-eyed vireo, most vireos live in undergrowth or thickets. They are not easy to see as they flit through the foliage, their drab colors doing little to distinguish them from any other small brown bird. Indeed, bird-watchers often find it easier to identify them by their voice or habits.

A few vireos have melodic songs, such as that of the warbling vireo, *V. gilvus*, of North America and the brown-capped vireo, *V. philadel-phicus*, of tropical America. The songs of these two birds are continuous warblings. The other vireos also are persistent songsters but there is little that is musical about them. The red-eyed vireo was once known as the preacher because of its persistent chattering. Each phrase consists of a variety of half a dozen notes, and one indefatigable ornithologist counted 22,917 phrases from one red-eyed vireo in a day, an average of over almost 1,000 an hour. The white-eyed vireo, *V. griseus*, varies its collection of clicks and mews with mimicked notes of other birds. Its song has

Vireos generally have a dull plumage and keep out of sight in forest or scrubland. As a result, sightings are rare and they may be identified more readily by their song than by sightings.

The typical vireo nest is suspended from a fork in a tree. It usually comprises an outer layer of strips of leaves and bark bound by spiderweb and lined with fine grass stems.

RED-EYED VIREO

CLASS	**Aves**
ORDER	**Passeriformes**
FAMILY	**Vireonidae**
GENUS AND SPECIES	***Vireo olivaceus***

WEIGHT
⁹⁄₁₀ oz. (17 g)

LENGTH
5 in. (12.5 cm)

DISTINCTIVE FEATURES
Dull gray crown and nape; olive-brown upperparts; pale buff underparts; small, robust bill; gray forehead; gray crown; broad white or grayish white supercilium (line above eye), bordered black above; red iris; dusky wings and tail; persistent, robinlike song

DIET
Insects; arthropods; small fruits

BREEDING
Age at first breeding: 1 year; breeding season: eggs laid April–June; number of eggs: 3 or 4; incubation period: 12–14 days; fledging period: 14 days; breeding interval: 1 year

LIFE SPAN
Not known

HABITAT
Mainly forest canopy; forest edge; open woodland; wooded agricultural land; suburbs

DISTRIBUTION
Much of North America except far north and southwest; South America, from Colombia south to northern Argentina

STATUS
Common

sometimes been described as sounding like "hick-whiskey-beer-hick," while the song of the blue-headed vireo, *V. solitaricius,* has been described as "see you, cheerio, be seein' you, so long, see ya." These interpretations serve as useful pointers to identifying birds that cannot be seen due to the fact that they are singing deep in undergrowth.

Many of the species of North American vireos are migratory, spending the winter in Central America. The red-eyed vireo migrates across the Gulf of Mexico, down Central America and into South America as far as southern Brazil.

Foliage searchers

Vireos feed mainly on insects and their larvae, spiders and a few small fruits. They forage for their prey among the foliage, which the vireos move through with agility, sometimes hanging upside down to search the undersides of leaves. They hold down large insects with one foot and attack them by thrusting and tearing with the bill. Only rarely do vireos search for food on the ground, although red-eyed vireos sometimes descend to feed on small snails.

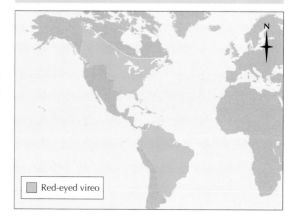

Red-eyed vireo

Singing on the nest

The nest is built in a tree or bush, sometimes near the ground but never on it. It is a deep cup of grass, leaves, strips of bark and other materials and is slung in a horizontal fork in a tree branch, supported by cobwebs. The female does most of the building and lays a clutch of three or four eggs in temperate regions or two or three eggs in the Tropics. They are white or cream with brown or lilac spots. Incubation lasts 12–14 days and the male helps, although he sits on the eggs for a shorter time than the female does. The males are such dedicated singers that they even sing while sitting on the nest. Not all males incubate, however, and in those species in which males do not share incubation duties they do not brood the young either. Otherwise, the urge to incubate is very strong and vireos sometimes have to be lifted off the nest in order to count their eggs. They may even peck at the offending hand. The chicks are fed by both parents, the female bringing most of the food, and they start to fly when they are about 2 weeks old.

Transatlantic immigrants

Although their distribution is American, red-eyed vireos may be seen in Britain, especially on the western coast of Ireland and on the Isles of Scilly, off the southwestern tip of England, in late September or October. Eleven were recorded in the United Kingdom in 1995. The red-eyed vireo is one of a growing band of American birds that have been seen in Britain. These include the pectoral sandpiper, *Calidris melanotos*, various ducks and cuckoos, the American robin, *Turdus migratorius*, the bobolink, *Dolichonyx oryzivorus*, and the Baltimore oriole, *Icterus galbula*.

It might seem understandable for shorebirds and waterbirds to cross the Atlantic, but many transatlantic migrants are small passerines. Some ornithologists have suggested that, having been blown out to sea, these small birds hitch lifts on ships for the rest of the way. However, they could not survive for such a period without food, many being insect-eaters, and if such a form of hitchhiking were the answer, there should be as many records of European birds reaching North America. Red-eyed vireos and other North American migrants set off in a generally southerly direction. Some follow a direct route south that takes them over the western part of the Atlantic Ocean. They are, in effect, taking a shortcut on their way to the Caribbean or South America. Some are caught up in fast-moving depressions sweeping west across the North Atlantic. First landfall for those that do not get swept into the ocean on the way is likely to be Ireland, the United Kingdom or Iceland. The red-eyed vireo is one of the most common passerines to cross the Atlantic, though few make it.

The female red-eyed vireo below is being fed by the male while she incubates the eggs. When the female leaves the nest the male takes over incubation.

VISCACHA

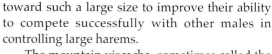

THE PLAINS VISCACHA, *Lagostomus maximus*, and the mountain viscacha, *Lagidium peruanum*, are South American rodents closely related to the chinchillas (discussed elsewhere in this encyclopedia). The two viscacha species are different in both form and habitat. The plains viscacha is heavily built; an adult male may weigh up to 18 pounds (8 kg). The body is up to 26 inches (66 cm) long with a short 6–8-inch (15–20-cm) tail. The head is large and blunt with prominent black whiskers. The tail, which is fully furred, is short and stiff and helps the animal to sit upright. The four fingers on each forefoot are well developed for burrowing and the three toes on each hind foot have extremely sharp claws. The fur is coarse and fairly long, dark gray above with white underparts and black-and-white stripes on the face. The female has somewhat lighter fur than the male and is considerably smaller. Indeed, the male may weigh up to four times more than the female. Male viscachas tend toward such a large size to improve their ability to compete successfully with other males in controlling large harems.

The mountain viscacha, sometimes called the mountain chinchilla, looks more like a chinchilla or a long-tailed rabbit. It is smaller than its plains relative, 12½–15¾ inches (31–39 cm) long in head and body, with a long tail of up to 12½ inches. It is a much slighter animal, weighing up to only 3½ pounds (1.6 kg). It has large, erect ears and, unlike the plains viscacha, the claws on the fore- and hind feet are blunt and weak. The fur is short, thick and soft. The upperparts vary in color from fawn to dark gray and the underparts are whitish, yellow or light gray. There is a crest of stiff hairs on the tail, which is black to reddish brown, and it often has a black stripe down the back.

Today there is only one species of plains viscacha, found over most of Argentina and north almost to southern Brazil. Another species, *L. crassus*, is now almost certainly extinct. The three subspecies of mountain viscachas are found in the Andes and its foothills, up to an altitude of as much as 17,000 feet (5,100 m), from Peru and Bolivia southward to northern Chile.

Social animals

Both the plains viscacha and mountain viscacha live in colonies, the former on the pampas and scrubland and the latter in rugged, mountainous country wherever water and food are available. The plains viscacha digs extensive burrows with long tunnels and numerous entrances, known as viscacheras. Some have been in continual use for centuries and sometimes cover as much as 6,455 square feet (600 sq m). Colonies of 15 to 30 individuals usually are formed, ruled over by a single adult male. Viscachas are very clean rodents, carrying all their refuse out of the burrow to pile on the excavated earth at the entrance. The entire surrounding area is cleared of brush, but they seem to have a passion for collecting objects and adorning the mounds with them. Stones, bones, cow dung, branches and even objects accidentally dropped by humans are dragged to the burrows and placed on top of the mounds.

The plains viscacha is nocturnal, coming out to feed in the evening. It shares its burrows with many other

Viscachas are large, heavyweight rodents that often sit in an upright position. They also have long, mustachelike whiskers.

PLAINS VISCACHA

CLASS	**Mammalia**
ORDER	**Rodentia**
FAMILY	**Chinchillidae**
GENUS AND SPECIES	***Lagostomus maximus***

WEIGHT
Male: up to 18 lb. (8 kg).
Female: 4⁹⁄₁₀ lb. (2 kg).

LENGTH
Head and body: 18–26 in. (47–66 cm);
tail: 6–8 in. (15–20 cm)

DISTINCTIVE FEATURES
Large, blunt head; mustachelike whiskers;
black-and-white stripes on face; 4 digits
with sharp, stout claws on forefeet; pad of
stiff bristles on hind feet; fur-covered tail;
coat coloration varies according to locality

DIET
Seeds and grasses

BREEDING
Age at first breeding: 8½ months (female),
15 months (male); breeding season:
March–April; gestation period: 150–155
days; number of young: usually 2; breeding
interval: 1 year (northern Argentina) or 6
months in favorable conditions

LIFE SPAN
Not known

HABITAT
Pampas (grassland) and lowland desert scrub

DISTRIBUTION
Southwestern Paraguay; southern Bolivia;
northern and central Argentina

STATUS
At low risk; populations declining in
Argentina due to hunting and eradication
programs

Plains viscacha

creatures: owls, snakes, lizards and even skunks. It also is friendly with its own kind, and occupants of neighboring colonies visit each other during the night. Each year's young males disperse from their birth colony and compete to gain entry to another colony. Previous resident males generally give way to challenging immigrant males. This is probably an adaptation to encourage out-breeding among colonies. By contrast, females are long-term residents in the colony in which they are born.

The plains viscacha employs a variety of calls from grunts and squeals to a wire-twanging sound. The warning note is a peculiar swishing noise followed by a liquid note that sounds like a drop of water falling into a pool.

The mountain viscachas have a very different way of life to that of the plains viscachas. Although they live in colonies comprising up to 80 individuals, they do not burrow but shelter in

Vegetation is sparse in the mountain viscacha's habitat. The animal feeds on most plants, but its diet comprises mainly mosses, grasses and lichens.

rock crevices or among piles of boulders. Unlike their relatives on the plains, they are diurnal and spend most of the day basking in the sun. They feed in the evenings but always return to shelter before dark. They are very agile, running among the rocks and leaping up the mountainside with their long hind legs.

Grassland inhabitants

The plains viscachas feed on a wide variety of grasses, roots, stems and seeds. They are regular feeders, often laying bare large areas of grassland. Some naturalists believe that 10 plains viscachas can eat as much grass as one sheep. In captivity they also eat carrots and potatoes. The mountain viscachas feed on plants, including grasses, mosses and lichens found near their colonies. Viscachas compete with domestic animals for pasture and it is possible that their acidic urine may damage grass for some time, factors that have caused them to be regarded as agricultural pests by farmers.

Slow breeding

The plains viscacha is a slow breeder. After a gestation of slightly less than 5 months, two young are born in July–August. There is usually only one litter a year, though if conditions are favorable there may be two, and the young do not reach maturity for 2 years. The mountain viscacha mates in October and November, and usually only a single young is born after a gestation of about 17–20 weeks, but there may be two or three litters a year. The young are able to nibble plants an hour after birth, and males become sexually mature in 7 months.

Plains viscacha under threat

Now that the puma, *Puma concolor*, has disappeared from much of its range, the plains viscacha has almost no predator except humans. As a result of the wide devastation of areas by the rodent's burrowing and its voracious appetite for grass, it is now regarded as a serious pest and in the past few years it has been cleared from many areas of Argentina, especially where grazing animals are kept. Viscacha meat is not eaten by local people, although when canned viscacha was exported to Italy in the 20th century, it proved to be popular. Viscacha fur is not valuable, but it is still used by locals and exported.

By contrast, the mountain viscacha is hunted by local people for food and for its hair, which can be mixed with wool and made into a yarn. It also has a natural predator in the Andean or culpeo fox, *Pseudolopex culpaeus*, and the Andean mountain cat, *Oreailurus jacobitus*. Both viscacha species are now sparsely distributed throughout their range, and the mountain viscacha, in particular, seems in some danger of extinction.

The mountain viscacha inhabits rugged montane country and relies on its thick, soft pelage for warmth. This viscacha was photographed in the High Andes, Peru.

VIVIPAROUS LIZARD

THE VIVIPAROUS LIZARD of Europe and Asia has a slender body with a long, tapering tail and well-developed limbs; a short, flat tongue, not deeply forked; and external ear openings. The two halves of the lower jaw are firmly connected. The male averages 5 inches (12.5 cm) in length, the female about 7 inches (17.5 cm). The female is the more heavily built of the two sexes, with the male the more slender, his tail tapering gradually to a fine tip. The tail is equal in length to the head and body in both sexes, but that of the female appears shorter because it tapers abruptly beyond the thick basal portion.

Viviparous lizards are very variable in color and pattern. The basic color is usually a shade of brown, but it may be olive, pale green or gray. Males usually have a pattern of black or brown spots and markings interspersed with cream-colored spots of equal size. They are yellow or orange on the underside, speckled with black markings. The dark markings on the back of females are less likely to have cream spots. They frequently coalesce, sometimes to form dark stripes that run along the middle of the back and along the sides of the back, when they may be edged by a pale stripe. The underside of the females is usually pale brown.

Viviparous lizards range farther from the equator than any other reptile, Varangerfjord in Norway, at 70° N, and Karesuando in Sweden, at 68° 30′ N, marking the most northerly limits. They extend south to northeast Spain, northern Italy and the Balkans, and eastward across Asia to Sakhalin Island in the China Sea. They are also the only reptiles that occur in Ireland. Viviparous lizards occupy a larger area than any other terrestrial reptile, and may be the most abundant reptile on Earth. At one time naturalists believed individual viviparous lizards in the Pyrenees laid eggs, but these are now regarded as a separate species.

Variety of habitat

The viviparous lizard, found in a wide variety of climates and habitats, is one of the hardiest reptiles. It lives as high as 8,000 feet (2,500 m) in the Balkans and extends north of the Arctic Circle in Lapland. Elsewhere it is common on heaths, in open woods, in hedgerows, in gardens and on sand dunes. It often basks in the sun and individuals have favorite sunning spots, such as a patch of sand or on an old wall. As many as 50 have been seen lying together, basking, with their bodies flattened and limbs extended to catch as much sun as possible. However, the viviparous lizard is intolerant of excessive heat and in southern Europe is found only in mountain districts, where it can keep cool.

The viviparous lizard runs with a nimble glide, shooting forward in short dashes from one tuft of herbage to the next, the body and tail scarcely lifted from the ground. It also can run easily over the tops of heather shoots, spreading its toes to cover gaps between the foliage.

In Britain, hibernation begins in October, the adults retiring before the young which, in a warm fall, are visible in southern England as late as November. The viviparous lizard is one of the first reptiles to reappear in the spring. To the south of the range this may occur in February, but usually it is in March, the males and young emerging first and the females some weeks later.

Fond of spiders

Viviparous lizards feed mainly on insects, including flies, beetles and moths, as well as ants and their larvae, and they are particularly fond of spiders. They swallow small caterpillars whole, chewing large ones, swallowing the insides and rejecting the skin.

This pregnant viviparous lizard will retain her eggs for about 3 months, until they are ready to hatch. The term viviparous means "giving birth to live young."

The viviparous lizard uses its claws to ascend sharp inclines. It also can swim well enough to pursue prey in water.

VIVIPAROUS LIZARD

CLASS	**Reptilia**
ORDER	**Squamata**
SUBORDER	**Sauria**
FAMILY	**Lacertidae**
GENUS AND SPECIES	***Lacerta vivipara***

ALTERNATIVE NAME
Common lizard (UK)

LENGTH
Usually up to 7 in. (18 cm)

DISTINCTIVE FEATURES
Small, slender body; brownish color; long, tapering tail; short, flat tongue; external ear openings

DIET
Invertebrates, especially spiders

BREEDING
Breeding season: July–August; number of young: 4 to 10; gestation period: 90 days

LIFE SPAN
Up to 10 years, but usually far less

HABITAT
Very varied, including Arctic tundra; heaths; moorland; roadside verges; cliffs; scrub; open deciduous woodland

DISTRIBUTION
Northern Norway and Sweden south to northeast Spain and east to Sakhalin Island in China Sea

STATUS
Abundant in some areas

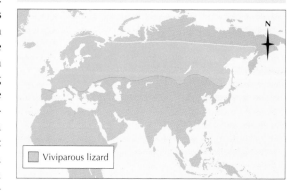

Viviparous lizard

The viviparous lizard locates insects by sound and may spend several minutes looking for one after hearing the rustle of the insect among grass or dry leaves. In captivity, it has been seen to enter a large bowl of water to seize an insect that had fallen onto the surface.

Living young

Many species of lizards and some snakes retain their young until they are fully developed. It is the secret of success at surviving in cool climates. When she is full of eggs, the female viviparous lizard spends as much time as possible lying in the sunshine. This means that the eggs are warmer than they would be if they were laid in the soil, and so they develop rapidly. The young are born free from the egg membrane, or else the membrane is broken either during or immediately after leaving her body. When kept in captivity, the mother makes no attempt at a nest or concealment and seems to take no interest in her young. In the wild, however, she digs a shallow pit, preferably well concealed in moist soil, into which she deposits her young in July or August. There are 5 to 8 in a litter, exceptionally 4 or 10. The young lizards are 1½–2 inches (3.8–5 cm) long at birth. Most of them are bronze brown but a few are born black and change to bronze brown within a week. The underparts are grayish brown and the back and sides are often speckled with gold. The young begin to feed within a few hours, hunting small soft-bodied insects such as aphids. From birth they are agile and skillful in the search for food. Male viviparous lizards reach sexual maturity at 21 months.

VULTURES

THE NAME VULTURE ORIGINALLY was applied to only the large, scavenging birds of the Old World that belong to the family Accipitridae, but after the first European settlers arrived in North America the term was extended to include the condors, turkey vultures and other members of the New World family Cathartidae. American vultures resemble the Old World vultures in appearance, presumably as a result of convergent evolution, and both groups have similar habits.

Classification

Vultures have a long fossil history, dating back to the Lower Miocene, over 20 million years ago. The Old World vultures are members of the same family as the hawks, eagles, falcons and relatives. Their closest relatives are the seven species of fish eagles in the genus *Haliaeetus* and the snake-eating

A flock of African white-backed vultures, Gyps africanus, gathers to feast on the remains of a large mammal carcass.

eagles of the genus *Cicaetus*. The New World vultures, which evolved separately and more recently than the vultures of Europe, Africa and Asia, are most closely related to the storks (order Ciconiiformes). There are seven species of New World vultures, including two species known as condors, and 14 species of Old World vultures.

Physical adaptations

Vultures have naked or nearly naked heads, necks, legs and feet, which is an asset to birds that regularly thrust their heads into, and stand on top of, rotting carcasses. The featherless areas are easy to keep clean, preventing bacterial and fungal growth. Unlike the true raptors (birds of prey), which kill their food, vultures have relatively weak feet. Vultures' feet are adapted for running rather than for holding prey.

Both families of vultures have heavy bodies, but despite this they can soar effortlessly for hours on their long, broad wings. The larger species are unable to sustain a flapping flight for more than a few seconds at a time. Vultures are, therefore, dependent on finding rising air currents in order to gain altitude. In steppes, plains and other flat country they use warm bubbles of rising air called thermals, and cannot fly until the heat of the sun has started to heat up the air. As a result, the vultures are effectively grounded during the early morning. However, vultures that live in hilly or mountainous regions, including both condors, can glide on the powerful updrafts that are created by air being forced up and over the undulating terrain.

Cathartid (New World) vultures differ from the other species primarily in anatomical characteristics, such as their perforated nostrils. It is possible to see right through the nostril opening of a turkey vulture, *Cathartes aura*, or a black vulture, *Coragyps atratus*.

Variety of species

Old World vultures have dark brown or black plumage, except in a few cases. The bare skin of the head and neck may, however, be pink, orange or white. The cinereous or Eurasian black vulture, *Aegypius monachus*, is one of the largest flying birds in the Old World. It has a wingspan of over 8 feet (2.45 m) and weighs more than 15 pounds (6.8 kg).

A pair of black vultures scavenging on the carcass of a bobcat, Everglades National Park, Florida.

New World Vultures Family Tree

ORDER	**Cathartiformes** **New World vultures**				
FAMILY	**Cathartidae**				
GENUS	**Vultur**	**Gymnogyps**	**Sarcorhamphus**	**Coragyps**	**Cathartes**
SPECIES	*Andean condor*	*California condor*	*King vulture*	*Black vulture*	*Turkey vulture* *Greater yellow-headed vulture* *Lesser yellow-headed vulture*

Its plumage is almost wholly dark brown or black, with pale skin on the head and neck. The cinereous vulture ranges from Spain east to Korea and Japan. At the other end of the scale there is the lammergeier or bearded vulture, *Gypaetus barbatus*, and the Egyptian vulture, *Neophron percnopterus*. The latter species has a wingspan of more than 5 feet (1.5 m) and is almost pure white except for black on the wings. The Egyptian

Old World Vultures Family Tree

Falconiformes
Old World vultures and relatives

Accipitridae

...pohierax	Gyps	Pseudogyps	Sarcogyps	Aegypius	Torgos	Trigonoceps	Gypaetus	Necrosyrtes	Neophron
Palm nut vulture	Griffon vulture	African white-backed vulture	Indian black vulture	Cinereous (Eurasian black) vulture	Lappet-faced vulture	White-headed vulture	Lammergeier or bearded vulture	Hooded vulture	Egyptian vulture
	Himalayan griffon vulture	Indian white-backed vulture							
	Rüppell's vulture								
	Long-billed vulture								

vulture ranges through southern Europe, Africa, the Middle East and India. Only a little larger is the hooded vulture, *Necrosyrtes monachus*, which is dark brown with a pinkish head and neck. It is common in sub-Saharan Africa.

The seven species of griffon and white-backed vultures (genera *Gyps*, *Pseudogyps* and *Trigonoceps*) are perhaps the typical vultures. They are found throughout southern Europe, Africa and Asia, often in large groups, and they nest in colonies. They are medium-sized and have a ruff of long feathers around the naked neck. The remaining Old World vultures include the palm nut vulture, *Gypohierax angolensis*, which has a feathered neck and striking black-and-white plumage, and the Indian black vulture, *Sarcogyps calvus*, which has a bright red head and neck.

The turkey and black vultures of the New World are relatively nondescript birds. The former has brown upperparts and a blackish body, with a naked red head (brownish in young birds). As their name suggests, the

two species of yellow-headed vultures (genus *Cathartes*) have bright yellow skin on their heads. The king vulture, *Sarcorhamphus papa*, has pure white plumage, black on the tail and wings, and colorful skin on the face. Apart from its huge size, the Andean condor, *Vultur gryphus*, is notable for the fleshy, comb-like caruncles that adorn the heads of adult males. The Andean condor and the California condor, *Gymnogyps californianus*, are the largest flying birds in the world. Their wingspans reach over 10 feet (3 m).

Vultures tend to have bare heads, necks, legs and feet. Being featherless, these areas are easier to keep dry and free of blood when feeding on carcasses. The king vulture (right) has the most colorful bare patches of any vulture.

Egyptian vultures are able to smash the tough shells of ostrich eggs by throwing a stone at them. Breaking an egg usually takes many attempts, but the nutritious contents are an ample reward. This is one of the few examples of tool use in birds.

Specialists at scavenging

Vultures hunt by sight, detecting carrion from vast distances by watching the behavior of other vultures and carrion-eating mammals such as hyenas and jackals. Big carcasses may attract large flocks of vultures, but despite their heavy bills most species have difficulty breaking through the skins of large animals. Therefore they have to wait for the carcass to decompose or for another animal to attack it. The large vultures, such as the lappet-faced vulture, *Torgos tracheliotus*, are powerful enough to rip through hide, and, although solitary in habits, they take precedence over the gregarious griffon and white-backed vultures at a carcass. These species, in turn, keep away the small species such as the Egyptian and hooded vultures, which have to be content with scraps.

The rasplike tongues of vultures enable them to pull flesh into their mouths, and their long necks allow them to probe deep into a large carcass. Vultures do not feed on carrion exclusively, however. The largest vultures occasionally prey on the chicks of flamingos or on small rodents. The palm nut vulture feeds on oil palm nuts as well as shellfish from the seashore, and sometimes hunts in shallow water for small fish.

Huge nests

Unlike the condors and some true birds of prey, the Old World vultures build their own nests instead of laying their eggs on the ground or in the abandoned nests of other birds. The lammergeier and the Egyptian vulture nest in caves or rock crevices, as do the griffon vultures, which nest in colonies of over 100 on cliffs. The white-backed vultures often nest in trees, with up to a dozen nests in one large tree. The larger vultures, the hooded vulture and the palm nut vulture, all nest singly in trees. The nests are huge cups of sticks and twigs lined with leaves, pieces of hide and refuse.

There usually is a single egg, two in smaller species, which is incubated by the female. Incubation ranges from 46 to 53 days, depending on the size of the vulture, and the chicks stay in the nest for up to 4½ months. The male feeds the female while she is incubating, and then both parents feed the chicks by regurgitation.

Decreasing in number

Vulture numbers are decreasing wherever intensive agricultural methods and modern hygiene are being introduced. There are fewer carcasses left lying in the open, and those that remain often have been poisoned by farmers who regard vultures and birds of prey as a nuisance. Although the vultures are not so useful nowadays as scavengers around human settlements, they still help to clear up the carcasses of stock, which are a potential source of infection. Unfortunately, they are not always seen in this light and are persecuted for allegedly killing livestock, although only the largest vultures could possibly attempt to do so.

Riding the thermals

Vultures are most common in dry, open country, and are also found in mountains up to 20,000 feet (6,100 m). As well as supplying the air currents necessary for flight, these areas also are places where there are likely to be plenty of carcasses of large animals easily visible from the air. Vultures are rarely found in forests, except for the hooded and king vultures. The hooded vulture is the most widespread, although not the commonest, vulture in Africa. It regularly scavenges around towns and villages, providing a valuable garbage disposal service, and even follows people as they till the soil, to feed on insects that are turned up. Because of its exploitation of humans, it is able to penetrate forests where there are human settlements.

Tool users

There are very few animals that use tools—the woodpecker-finch (one of the Darwin's finches), the chimpanzee and the sea otter are the best known. In 1966 another was added to the list. This is the Egyptian vulture, which throws stones at eggs. It smashes the tough shells of ostrich eggs either by throwing them against a rock or another egg or by throwing a stone at them. If there is no stone nearby, a vulture may search for one up to 50 yards (46 m) away, fly back with it in its bill and then sling it with a violent downward movement of the head.

For particular species see:
• CONDOR • LAMMERGEIER • TURKEY VULTURE

WAGTAIL

WAGTAILS ARE SMALL BIRDS that are closely related to pipits but have brighter plumage and characteristic long tails that continually wag up and down. The bill is needlelike, typical of insect-eaters, and the feet are well developed, with long toes. The tail is nearly as long as the head and body, making a total length of 6–7½ inches (15–19 cm).

There are 12 species of wagtails, some of which are divided into races with separate common names. In the case of the yellow wagtail (*Motacilla flava*), for example, the race that lives in Britain and on the nearest parts of the Continent has more yellow in its plumage than the race that breeds farther east and is known as the blue-headed wagtail. Both races are greenish brown above and yellow underneath, with white outer tail feathers. The male of the British yellow wagtail has a bright yellow crown and eye stripe, whereas the male of the continental race has a slate-blue crown, white eye stripe and white chin. The females of both races are similar to each other but have the same eye stripes as the respective males. Ornithologists consider that there are 18 subspecies, or races, of the yellow wagtail, although there is much debate on the subject and a number of these races are now regarded by some as separate species.

The gray wagtail, *M. cinerea*, could easily be mistaken for a yellow wagtail because its underparts are yellow, but it can be distinguished by its blue-gray upperparts. The male has a black bib in summer. Closely related to the gray wagtail are the mountain wagtail, *M. clara*, of Africa and the Madagascar wagtail, *M. flaviventris*. The pied wagtail, *M. alba*, has eight races, which are easily confused. Of the two that are seen in Britain, both have distinctive black-and-white plumage, but in summer one of them, the white wagtail, *M. a. yarrellii*, a vagrant from continental Europe, has a light gray back. The Cape wagtail, *M. capensis*, of southern Africa is a close relative.

Up and down and side to side

Wagtails live in Europe, Asia and Africa and are occasional visitors to Australia. The yellow wagtail has crossed the Bering Strait and breeds in western Alaska, while the white wagtail has bred in Greenland. The forest wagtail, *Dendronanthus indicus*, lives in Manchuria and Korea and is rather different from other wagtails: it wags its tail from side to side rather than up and down. The willie wagtail, *Rhipidura leucophrys*, of Australia is not a true wagtail. It is a fantail (family Muscicapidae) but also has the habit of wagging the tail.

Motacilla flava flavissima, *which breeds in Britain, is one of 18 subspecies of yellow wagtail. It sometimes interbreeds with M. flava flava, *the blue-headed wagtail.

A male blue-headed wagtail sings to advertise his ownership of a territory. His song acts as a warning to rivals and an invitation to females.

AFRICAN PIED WAGTAIL

CLASS	**Aves**
ORDER	**Passeriformes**
FAMILY	**Motacillidae**
GENUS AND SPECIES	*Motacilla aguimp*

WEIGHT
¾–1¼ oz. (22–33 g)

LENGTH
Head to tail: 7¼–7½ in. (18.5–19 cm)

DISTINCTIVE FEATURES
Long tail; black upperparts; white outer tail feathers; white wing patch; white supercilium (stripe) in front of and behind eye; white throat and belly; black gorget on breast

DIET
Adult insects and larvae; domestic scraps

BREEDING
Age at first breeding: 1 year; breeding season: year-round, primarily in wet season; number of eggs: 3 to 4; incubation period: 13–14 days; fledging period: 15–16 days; breeding interval: 1 year

LIFE SPAN
Up to 9 years

HABITAT
Often along rivers, on sandbanks; also by coastal lagoons; on fields, playing fields near towns; mostly in lowlands

DISTRIBUTION
Very discontinuous in sub-Saharan Africa

STATUS
Common in good habitat

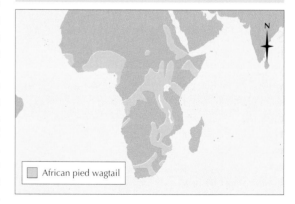

African pied wagtail

Wagtails usually live in open country, particularly in grasslands, but the pied wagtail is sometimes found in trees and often occurs on farms and around houses, while the forest wagtail lives in woods. The pied and gray wagtails, notably the latter, are often found near water, sometimes wading into it. Many wagtails migrate, especially the yellow wagtail, the European population of which winters in Africa. The species also occasionally reaches Australia from Asia. In winter, many wagtails roost communally among dense vegetation or in trees. Pied wagtails sometimes roost in the eaves of buildings, which may contain hundreds of birds. In spring, yellow wagtails gather in huge reed-bed roosts prior to returning to Eurasia. One roost in Queen Elizabeth National Park, Uganda, sometimes holds 2 to 3 million birds in March.

Bounding flight

The flight of wagtails consists of a series of arcs as they alternately beat and close their wings. The glides between each burst of beating are longer than those in the flight of finches, and consequently the flight of wagtails is much more undulating. Their gait on the ground also is very characteristic. They run swiftly after insects on their strong legs, with head, body and tail parallel to the ground, and then stop suddenly

and bob their tails rapidly. When a wagtail is walking normally, the tail bobs and the head nods in time to the step.

Wagtails eat mainly small insects, particularly flies and insects living on the ground. Wagtails living near water feed on water insects such as small dragonflies and water beetles, small snails and even minnows. Most food is caught either on the ground or just above it, but wagtails also can be seen fluttering up to catch a higher-flying insect then dropping back to the ground. Yellow wagtails are found with cattle or sheep, feeding on the insects that they disturb.

Sitting on their tails

In his book *The Yellow Wagtail*, Stuart Smith gave vivid descriptions of the rivalry between male yellow wagtails at the beginning of the breeding season. Each male stakes out a territory and defends it against neighbors or other males that try to take it over. The owner advertises his possession by a warbling song and displays to rivals by throwing his head back, puffing out his breast feathers and leaning back on his tail so it is flattened against the ground. Occasionally fights break out on the ground or in the air and rivals peck and claw each other.

Once territorial ownership has been settled and mating has taken place, the female builds a nest. The male does not assist but merely escorts her. The nest of grass lined with hair usually is

built among long grass or undergrowth. Pied and gray wagtails often nest in cavities in walls, banks or trees and sometimes use old birds' nests, such as those of blackbirds or dippers.

The clutch generally consists of three to six eggs, which are incubated for 2 weeks almost entirely by the female. The young are fed by both parents. They leave the nest after 2 weeks and continue to be fed, staying together as a family party for some time. The pied wagtail raises two or three broods a year, the male caring for one brood while the female incubates the next clutch.

Why wagtail?

The Saxon name for the wagtail was *wagstyrt*, *styrt* meaning tail, and its name in other languages also refers to the habit of bobbing the tail. In Dutch, the wagtail is *Kwikstaart* and in Danish *Vipstjert*. Tail-wagging is not confined to wagtails. It is well developed but not as obvious in the related pipits, for example.

The reason for tail-wagging is obscure. In wagtails it is probably related to the habit of suddenly dashing and fluttering after insects. The long tail acts as a counterpoise to the body and aids balance as the wagtail stops dead or changes direction. Other birds twitch their tails after landing. This wagging movement, of which the wagtails' bobbing may be an extreme development, could equate to the human characteristic of waving the arms when unbalanced.

Wagtails are insect-eaters, and the rapid dash of a pied wagtail across a piece of grass signifies the pursuit of a small prey item that the bird's keen eyesight has detected.

WALLABY

Wallabies are grazers and browsers. The Parma wallaby, Marcropus parma, *is found in forests in northeastern New South Wales, Australia.*

THERE ARE MANY different species of wallabies. Some, such as the quokka, rock wallaby and pademelon, have been dealt with elsewhere in this encyclopedia. In Australia the wallabies' story is largely one of persecution or extinction. There also are wallabies in New Guinea, and one species of wallaby has become acclimatized to living wild in parts of Europe.

Most wallabies are the size of a hare or slightly larger, but the brush or scrub wallabies may grow up to 3 feet (90 cm) long in head and body, with a tail 2½ feet (75 cm) long and a weight of up to 50 pounds (22.5 kg). There are three species of hare wallaby, each of which is grayish brown with some red in places. There is only one species of banded hare wallaby, *Lagostrophus fasciatus*. It is grayish with many dark bands across the back, from the nape running down to the base of the tail. The three species of nail-tailed wallabies are mainly gray with white stripes, but the 11 species of scrub wallaby are sandy to reddish brown in color. The five New Guinea forest wallabies are grayish brown to blackish brown.

Hares and organ-grinders

Hare wallabies are so named not only for their size but also for their habit of lying in a shallow trench, called a form, scratched in the ground under a bush or grass clump. They also run fast and have a tendency to double back on their tracks in a manner reminiscent of hares. Hare wallabies are solitary and nocturnal and make a whistling call when they are pursued. The nail-tailed wallaby, *Onychogalea fraenata*, has similar habits except that when bounding along it hold its small forelegs out to the sides and swings them with a rotary movement, which has earned it the nickname organ-grinder. The purpose of the spur or nail at the tip of the tail is not clear. Scrub wallabies spend much of the day hidden but often come out to feed, although they never stray far from cover. They may be solitary or live in pairs or in large groups. The New Guinea forest wallaby, *Dorcopsis veterum*, lives in the rain forest from sea level to 10,000 feet (3,000 m), but apart from this scientists currently have little information about it, and it is assumed that the species has a similar lifestyle to other wallabies.

WALLABIES

CLASS	**Mammalia**
ORDER	**Marsupialia**
FAMILY	**Macropodidae**
GENUS	**12**
SPECIES	**61**

WEIGHT
2²⁄₁₀–46²⁄₁₀ lb. (1–21 kg)

LENGTH
**Head and body: 12–36 in. (30–90 cm);
tail: 8⁸⁄₁₀–36 in. (22–90 cm)**

DISTINCTIVE FEATURES
**Small front legs; highly enlarged hind legs;
relatively large ears; long, thick tail; females
have pouch**

DIET
Grass and herbage

BREEDING
**(Swamp wallaby, *Wallabia bicolor*) Age at
first breeding: about 15 months; breeding
season: year-round; number of young: 1,
rarely 2; gestation period: 37 days if not
subject to delayed implantation, otherwise
young live in pouch for about 250 days;
breeding interval: 240 days**

LIFE SPAN
Up to 15 years

HABITAT
**Varied, including swamp, forest and rocky,
semiarid areas**

DISTRIBUTION
Australia and New Guinea

STATUS
**Many populations in decline due to
reduction in suitable habitat. Several species
hunted for meat or skin, or as an
agricultural pest.**

Wallabies

Generalized diets

All wallabies are vegetarian, most of them eating grass, although the nail-tails eat mainly the roots of coarse grasses and the scrub wallabies take succulent roots and eat leaves as well. As with other herbivores, the diets as given here probably are oversimplified. Some wallabies, normally grass eaters, have been seen to eat bark and fruits as well as leaves.

A variety of threats

The wallabies that are dealt with here are among the less fortunate species. They are small enough to be at the mercy of introduced foxes and dogs, the fur of some of these species is valued for export, their speed makes them targets for sport and their flesh is agreeable as human food. The aborigines of Australia and New Guinea have hunted them for food, some species, such as hare

Female wallabies usually give birth to a single young, or joey. Mother and offspring may stay together some time after the joey is weaned, which may be a year after birth.

wallabies, being persecuted more than others. The banded hare wallaby, now confined to a few islands in Shark Bay, southwestern Australia, once occurred in great numbers on the Australian mainland. At one time the aborigines burned the undergrowth to drive out the game, and this particular species of hare wallaby suffered badly as a result. The larger scrub wallabies have suffered from the general supposition that they are outgrazing the sheep and cattle, although recent research in Australia suggests this view may have been exaggerated with regard to all wallaby and kangaroo species. Scientists have known for some time that the reverse was true for the banded hare wallaby: it was displaced by grazing sheep.

Some species of wallabies have become extinct, whereas others are much reduced in numbers and range, and the only hope of ultimate survival for all seems to lie in creating suitable reserves and sanctuaries for them and generally improving the climate of public opinion in Australia regarding the species.

Fate of immigrants

Australia is not the only part of the world where wallabies have had a troubled existence. In his book *Four-legged Australians*, the German scientist Professor Bernard Grzimek summarized the experience of those Bennett's wallabies, *Macropus rufogriseus fruticus*, that were brought over to Europe. In 1887, the German nobleman Baron Philipp von Böselager released two males and three females in 250 acres (100 ha) of forest near Heimerzheim in western Germany. In spite of one hard winter, they multiplied to 35 or 40 animals in 6 years and the prospects for their survival looked promising, but subsequently the entire herd was exterminated by poachers. At the beginning of the 20th century, another nobleman, Count Witzleben, successfully bred Bennett's wallabies on his estate near Frankfurt-on-Oder. However, he later arranged for them all to be destroyed because he came to believe that they were alarming the deer that he kept. Prince Gerhard Blücher von Wahlstatt released wallabies on the island of Herm, in the Channel Islands, and they did well there until British troops occupied the island and, over time, consumed the entire herd. Today feral wallabies live on the moors in Derbyshire, central England.

One danger peculiar to temperate climates that the animals face comes from ice on lakes. This tends to break under the strain of a wallaby's heavy, rhythmical leaps, often with fatal consequences for the animal.

The bridled nail-tailed wallaby is a rare and endangered species. It is found only in Queensland, Australia and is now protected by law.

WALL LIZARD

Visitors to central and southern Europe cannot fail to notice wall lizards, which are abundant not only in the countryside but in towns and villages also. The fingers and toes are relatively long, and end in claws that enable the animals to run up and down vertical surfaces such as walls with grace and speed. This characteristic has given them their common name, but they are found in flat environments too. Adult wall lizards usually grow to about 8 inches (20 cm), but in some places, for example in central and southern Italy, they may be larger.

The species with the most widespread distribution is the common wall lizard, *Podarcis muralis*, which is often referred to simply as the wall lizard. This species is found from France and northern Spain eastward to Greece and the shores of the Black Sea. It is replaced in most of Spain and Portugal by the Iberian wall lizard, *P. hispanica*, and in southern Italy by the ruin lizard, *P. sicula*. In the Balkan Peninsula, there are seven species that can be regarded as wall lizards. Many Mediterranean islands have their own species of wall lizard. Malta, Sicily, Ibiza and Minorca have one each, for example, while Majorca has two, and Corsica and Sardinia share two more. Other species occur on some of the islands in the Aegean Sea. Wall lizards on small islands are sometimes larger than those on the mainland; they may be a different color, and sometimes they are all black. It is not surprising, therefore, that the classification of wall lizards has been a matter of controversy and sometimes heated debate among scientists and it is still not fully resolved. One of the problems faced by biologists in trying to determine how wall lizards have evolved is that it is not easy to establish how much their distribution has been modified by humans. The Phoenicians, Carthaginians, early Greeks and Romans, for example, all traded extensively within the Mediterranean basin, and wall lizards could easily have been accidentally carried from place to place as fugitives in cargoes of grain or timber.

Wall lizards occur in a wide array of colors and patterns. Males often have a mottled pattern of black and green; females tend to be mottled black and brown. Male wall lizards from Tuscany are usually bright green, while around Rome they are black with small green markings.

Benefit of global warming?

The northern limit of the distribution of wall lizards in Europe, like that of its relative, the larger green lizard, *Lacerta viridis*, follows approximately the limit of areas where wine-growing is successful and there are many vineyards. Wall lizards do not occur naturally in Britain. Several attempts have been made to introduce them, but until 20–30 years ago these had all ended in failure. Now, however, there are many wall lizard colonies in southern England, and most of them are expanding. The reason is that average temperatures have become slightly warmer and it is now possible for the eggs of the lizards to complete their development and hatch before the

The common wall lizard is the most widespread species in the genus Podarcis. *It is often visible climbing on trees, on rocks in open country and on buildings.*

Each wall lizard has a favorite spot on a rock or wall where it regularly basks. In cloudy or rainy weather, a wall lizard retires to crannies or holes in rocks.

onset of cooler weather in the fall. There are also several introduced colonies in Austria, some parts of Germany where they do not occur naturally, and in the United States, for example in Pennsylvania and Ohio.

Residents of ruins

Geckos and lizards, including many wall lizards, swarm among the archaeological and historical ruins that abound in southern Europe. In certain sites in southern Italy for example in the ruins at Pompeii, ruin lizards are abundant. Because so many people visit places of this kind, the lizards are used to humans and may become very tame.

Wall lizards are particularly abundant on many of the smaller islands and islets in the Mediterranean. Scientists are unsure as to why this is so. It may be that they face less competition from other animals in these areas, or there may be fewer predators. The lizards in these areas are often much larger than those on the mainland, and frequently their colors and patterns are different; there are a few small islands where the predominant color of lizards is blue. Lizards that are all black are also a feature of small islands, although the reason for this remains a subject of scientific debate. A few lizards on small islands are distinct enough to be described as separate species.

A further enigma is that in places where wall lizards are at the extreme northern edge of their range they only live on castles, forts and other buildings; they do not occur in the surrounding countryside. This is the case in an number of sites in Europe, for example just south of Bonn in Germany, Maastricht in the Netherlands, Jersey in the Channel Islands, (where their stronghold is Mont Orgueil Castle) and at the port of St-Malo on the adjacent French coast (where they are especially common on the ruins of the castle).

WALL LIZARDS

CLASS	**Reptilia**
ORDER	**Squamata**
SUBORDER	**Sauria**
FAMILY	**Lacertidae**

GENUS AND SPECIES **Common wall lizard, *Podarcis muralis*; Iberian wall lizard, *P. hispanica*; ruin lizard, *P. sicula*; others**

ALTERNATIVE NAME
Italian wall lizard (*P. sicula*)

LENGTH
(*P. muralis*) usually up to 8 in. (20 cm); up to 12 in. (30 cm) in central and southern Italy

DISTINCTIVE FEATURES
Slim body and tail; narrow, pointed head; long, slender toes and claws; scaly upperparts; color variable, including gray, reddish brown or black ground color with bronze or greenish tinge; blue, green or white spots on lower flanks; milky white or copper-red upperparts

DIET
Small insects; other small invertebrates; occasionally berries and small fruits

BREEDING
Breeding season: summer; 1 to 3 clutches

LIFE SPAN
Up to 10 years, usually much less

HABITAT
Varied, including gardens, parks, olive groves, waste ground, open shrubland and woodland

DISTRIBUTION
***P. muralis*: France and northern Spain eastward to Greece and Black Sea. Other species: widespread in central and southern Europe; nearly all islands in Mediterranean; land between Black Sea and Caspian Sea.**

STATUS
Abundant. Population estimated at 5,000 million in Italy alone in 1970s.

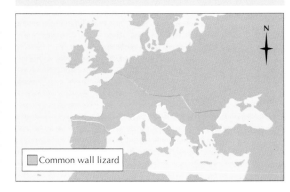

Common wall lizard

WALRUS

HUNTED SINCE THE time of the Vikings, sometimes almost to the point of extinction, the walrus has survived and today, with strict conservation measures, some herds are slowly recovering. The two walrus subspecies, the Pacific walrus and the Atlantic walrus, differ only in minor details. The Pacific bulls average 11–11½ feet (3.3–3.5 m) long and weigh slightly more than 2,000 pounds (900 kg) but can reach 13¾ feet (4.1 m) and weigh up to 3,700 pounds (1,665 kg) when they carry maximum blubber. The Atlantic bulls average 10 feet (3 m) long and up to 1,650 pounds (743 kg) in weight but may reach 12 feet (3.6 m) and weigh 2,800 pounds (1,260 kg). The cows of both subspecies are smaller, 8½–9½ feet (2.6–2.9 m) and 1,250 pounds (563 kg), but large Pacific cows may reach almost 12½ ft (3.8 m) and a weight of 1,750 pounds (788 kg).

The walrus is heavily built, adult bulls carrying sometimes 900 pounds (405 kg) of blubber in winter. The head and muzzle are broad and the neck is short, the muzzle being deeper in the Pacific walrus. The cheek teeth are few and simple in shape, but the upper canines are elongated to form large ivory tusks, which may reach 3 feet (0.9 m) in length and are even

longer in the Pacific subspecies. The nostrils in the Pacific subspecies are placed higher on the head. The mustache bristles are very conspicuous, especially at the corners of the mouth, where they may reach 4–5 inches (10–12.5 cm). The foreflippers are strong and oarlike, being about a quarter the length of the body. The hind flippers are about 6 inches (15 cm) shorter and very broad, but with little real power in them.

The walrus's skin is tough, wrinkled and covered with short hair, reddish brown or pink in bulls and brown in the cows. The hair becomes scanty after middle age, and old males may be almost hairless, with their hide in deep folds.

The Pacific walrus lives mainly in the waters adjacent to Alaska and the Chukchi Sea in Russia. The Alaskan herds migrate south in the fall into the Bering Sea and Bristol Bay to escape the encroaching Arctic ice, moving northward again in spring when it breaks up.

The Atlantic walrus is sparsely distributed from northern Arctic Canada eastward to western Greenland, with small isolated groups on the eastern Greenland coast, Spitsbergen, Franz Josef Land and the Barents and Kara Seas. It migrates southward for the winter.

Walruses are sociable animals and at one time regularly gathered on remote beaches. Due to persecution by hunters, however, they now tend to gather on less accessible ice floes.

The walrus uses its tusks for hauling itself along on the ice. The family name, Odobenidae, means "those that walk with their teeth."

Walruses also inhabit the Laptev Sea near Russia and do not migrate in the winter. Scientists believe this herd may be a race midway between the Atlantic and Pacific subspecies.

Sociable lifestyle

Walruses associate in family herds of cows, calves and young bulls of up to 100 individuals. Except in the breeding season the adult bulls usually form separate herds. They live mainly in shallow coastal waters, sheltering on isolated rocky coasts and islands or congregating on ice floes. As a result of persecution by humans, however, walruses avoid land as much as possible and to keep to the ice floes, sometimes far out to sea. They normally are timid but readily become aggressive in the face of danger. There seems to be intense devotion to the young, and the killing of her offspring will rouse a mother to a fighting fury, quickly joined by the rest of the herd.

Walruses can move overland as fast as a human can run, and because of their formidable tusks, hunters, having roused a herd, have often been hard put to keep them at bay. Walruses have even speared the sides of a boat with their tusks or hooked them over the gunwales.

As well as using its large tusks as weapons of offense and defense, the walrus uses them to keep breathing holes open in the ice. It also uses

WALRUS

CLASS	**Mammalia**
ORDER	**Pinnipedia**
FAMILY	**Odobenidae**
GENUS AND SPECIES	***Odobenus rosmarus***

WEIGHT
880–3,740 lb. (400–1,700 kg); male heavier than female

LENGTH
Head and body: 7¼–12 ft. (2.2–3.6 m); male larger than female

DISTINCTIVE FEATURES
Thickset, rotund body; thick, wrinkly skin; broad foreflippers; long tusks; mustache in both sexes

DIET
Crustaceans, especially clams and mussels; fish; rarely seals

BREEDING
Age at first breeding: 7 years (female), 15 years (male); male able to breed earlier but cannot compete with other males until 15 years old; breeding season: mainly January–February; number of young: 1; gestation period: about 14–16 months, including 4–5 months' delayed implantation; breeding interval: 1 calf every 2 years in most productive females

LIFE SPAN
Up to 40 years

HABITAT
Coastal waters in Arctic Ocean and adjoining seas; often found on pack ice

DISTRIBUTION
Arctic Ocean

STATUS
Threatened by hunting; some subspecies regarded as rare or vulnerable

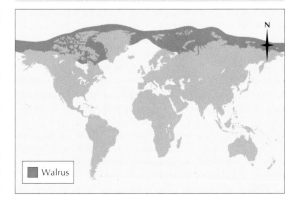

Walrus

them as anchors for hauling itself out onto the ice, heaving up to bring the foreflippers onto the surface. The horny casing of bare hard skin on the palms of the flippers prevents the walrus from slipping. Some scientists have suggested the tusks are also used to dig for food on the ocean floor, but there is no evidence to support this.

Walruses sunbathe, which turns their skin a pinkish color, and sleep packed close together on the ice floes with their tusks resting on each other's bodies. If the water is not too rough, adult walruses also can sleep vertically in the water by inflating the air sacs under their throats.

Clam grubbers

The walrus's diet consists principally of clams, which it grubs out of the mud with its muzzle, and sea snails. It also takes mussels and cockles. The snout bristles help in detecting the shellfish. In captivity a walrus was seen to suck out clams and discard the shells. A walrus also swallows a quantity of pebbles and stones, possibly to help it crush the food in its stomach. Walruses usually dive for their food in shallow water of about 180 feet (54 m) or less, but occasionally they go down to 300 feet (90 m). As yet, scientist are unsure how walruses deal with pressure problems at this depth, but they must have special physiological adaptations to do so.

Occasionally a walrus, usually an adult bull, will turn carnivorous and feed on whale carcasses or it may kill small ringed or bearded seals. Having sampled flesh, it may continue to eat it in preference to shellfish.

Hitchhiking pup

Most matings take place in January and February, and after a gestation of slightly more than a year one pup is born every alternate year. Birth takes place on an ice floe. The newborn pup is 4 feet (1.2 m) long with a coat of short silver gray hair and weighs 100–150 pounds (45–67.5 kg). It is able to swim immediately, although not very well, and follows its mother in the water. After a week or two it can swim and dive well. Even so, it usually rides on its mother's back for some time after birth, gripping with its flippers. After a month or two the silver gray hair is replaced by a sparser dark brown coat of stiff hairs. The cow nurses the pup for 18 months to two years, but they remain together for several months after weaning. Males become sexually mature at about 8–10 years, the females at about 6–7 years.

Slaughtered by hunters

Killer whales and polar bears attack walruses but not often, the polar bear particularly being wary of attacking an adult bull even when he is ashore and more vulnerable. Walruses have been hunted by humans from early times. Historically, the Inuit and Chukchi relied on the annual kill to supply all their major needs, including meat, blubber, oil, clothing, boat coverings and sled harnesses. Even today they are largely dependent on it. The annual killings by local people, however, had little effect on the numbers of the herds. It was the coming of commercially minded Europeans to the Arctic that started the real population decline. From the 15th century onward they used the walrus's habit of hauling out on the beaches in massed herds to massacre large numbers in the space of a few hours. After 1861, when whales had become scarce, whalers from New England started harpooning walruses. Then they began to use rifles and the Inuit followed suit. More walruses could be killed using a rifle, but many carcasses fell into the water and could not be recovered. An even greater wastage has been caused by ivory hunters, who kill for the tusks and discard the rest of the carcass. By the 1930s the world population of walruses stood at less than 100,000 and only recently have strict conservation measures been enforced. At present, both walrus subspecies are safe from threat.

Inflatable air sacs in their throats enable walruses to float while resting in the water. The sacs also enable them to make special sounds during courtship.

WAPITI

Call of the wild: a stag bellows a challenge to rivals during the rut, or breeding season, in which the adult males round up as many adult females as they can.

THE WAPITI IS THE NORTH AMERICAN form of *Cervus elaphus*, which exists in various subspecies across the whole of Europe and Asia. In Britain it is known as the red deer. In the 17th century, the wapiti was abundant and widely distributed in North America. Its range fell just short of the Pacific Coast in the west across almost to the Atlantic Coast in the east, and from British Columbia in the north to New Mexico and Arizona in the south. Its numbers then were about 10 million. The total today in North America is 500,000.

The wapiti is typically larger than the red deer, up to 8⅔ feet (2.65 m) long in head and body, 5 feet (1.5 m) high at the shoulder and weighing up to 750 pounds (340 kg), the hinds (adult females) being smaller than the stags (adult males). By comparison, red deer in Britain seldom measure more than 4 feet (1.2 m) high at the shoulder. The wapiti's antlers measure nearly 6 feet (1.8 m) across the beam. The wapiti resembles the red deer in color except that it is less reddish in summer and it has a more prominent pale rump patch. The name wapiti, given by the Native American Shawnee, means white deer and may refer to this pale patch.

Three major groups

Around the world, *C. elaphus* is broadly divided by some scientists into three major populations. The first is found in Europe, North Africa and Asia as far east as the shores of the Caspian Sea. Those in Europe number perhaps a million, the largest stocks being in Scotland and Germany. There are rare subspecies on Corsica and Sardinia in the Mediterranean (*C. e. corsicanus*), North Africa (*C. e. barbarus*) and the Caucasus and Asia Minor (*C. e. maral*).

The second population, which may be the ancient progenitor of all other *C. elaphus* stocks, is spread across Central Asia and north-central China. It comprises six geographically isolated subspecies: *C. e. bactrianus, yarkandensis, wallichi, hanglu, macneilli* and *kansuensis*. At least two of these are feared to be extinct due to uncontrolled hunting or habitat loss following the expansion of agriculture or cattle ranching.

The third group, which comprises a number of subspecies in Central and East Asia, also includes the wapiti. Of the American forms of wapiti the largest is found near the Pacific Coast. The smallest, the dwarf or tule wapiti, lives on the hot, dry plains of southern California. Its coat is much paler than those of the other three. The subspecies *C. elaphus canadensis*, once found across eastern-central North America, is now extinct, as is the southern and Mexican subspecies *C. e. merriami*.

In Britain, *C. elaphus* is marked out as the only deer in which the adult does not have a spotted coat. Also, the upper tine (point) on a British stag's brow is often absent. Those few stags that develop no antlers at all are known in Scotland as hummels and in southwestern England as notts. Red deer hybridize with the sika deer, *C. nippon*, which was introduced to

WAPITI

CLASS	**Mammalia**
ORDER	**Artiodactyla**
FAMILY	**Cervidae**
GENUS AND SPECIES	***Cervus elaphus***

ALTERNATIVE NAME
Red deer (Britain); elk (erroneously)

WEIGHT
165–750 lb. (75–340 kg)

LENGTH
**Head and body: 5½–8⅔ ft. (1.65–2.65 m);
shoulder height: 2½–5 ft. (75–150 cm);
tail: 4–11 in. (10–27 cm)**

DISTINCTIVE FEATURES
**Brown or chestnut upperparts; pale cream
to beige underparts; pale to white rump
patch; coat shaggier in cold climates; male
has manelike neck fur and large antlers**

DIET
**Leaves, shoots, bark, some grass, lichens,
tree shoots, heather, algae**

BREEDING
**Age at first breeding: female 28 months,
male 3–4 years; breeding season: September
to October; gestation period: 235–260 days;
number of young: 1 (rarely 2); breeding
interval: 1 year**

LIFE SPAN
Up to 20 years; 30 years in captivity

HABITAT
**Very varied, including moorland and
montane forest**

DISTRIBUTION
**Europe, western North America, Siberia,
China, Mongolia, northwest Africa;
introduced into Argentina, Chile, Australia,
New Zealand**

STATUS
Some subspecies threatened

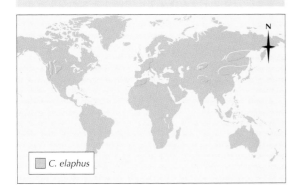

C. elaphus

Britain from eastern Asia in 1860–1920 and is increasing its range today in Scotland. The range of the British red deer has declined in size since prehistoric times, partly as a result of climate change but also because of human disturbance. In the wake of pressure from sport hunters, North American wapiti have been exported to British herds over the last century or so to improve the bloodline.

Wherever it is found on mainland Europe and Asia, *C. elaphus* is found in, or close to, woodland of almost any description, although larger stands of dense forest are rarely favored. In Scotland, herds are numerous on the open moors, browsing on heather, dwarf shrubs and lichens. Across their range red deer are highly versatile, eating anything from ferns and tree shoots to seaweed in coastal areas.

The calf can stand shortly after its birth, but if disturbed it tends to lie flat on the ground and await its mother's return. This 10-day-old calf is already steady on its hooves.

Separate lives

For most of the year the sexes live apart. Relatively stable female groups form around an alpha (dominant) female and her offspring. Local food resources affect not only the home range of the group but also its number. In favorable sites, groups (numbering 10 to 40 in Britain) may coalesce to form herds of several hundred, whereas herds within woodland tend to disperse into family units. Stags live in looser herds and occupy land of poorer quality. There is a hierarchy in each group, regardless of sex; among stags this is based on size and age. Wapiti are active at any time of the day or night, feeding and resting in up to nine cycles during 24 hours. They become more nocturnal, however, where disturbed. Those on hilly terrain tend to migrate uphill by day, returning to lower slopes or valleys at night.

Rallying cry

The breeding habits of wapiti are very similar to those of red deer. At the beginning of the rut, the wapiti stag gives an undulating bugling call, which starts in a low key, rises to a high pitch in a prolonged note that abruptly drops to a harsh scream, and ends in a few grunts. He also gives a loud, far-carrying roar. The mature bulls round up hinds into harems in September and October, to the accompaniment of a clatter of antlers and a clashing of foreheads from furious fighting, as the subordinate stags challenge their seniors.

National parks are vital lifelines for endangered subspecies of red deer. This stag, a resident of Yellowstone in the Rockies, has grizzly bears and, since 1995, wolves for neighbors.

In May or June of the following year, after a gestation period of 235–260 days, the hinds leave the herds and go into thickets to drop their dappled calves. There usually is one at a birth, or occasionally twins. The calf weighs up to 30 pounds (13.5 kg) at birth, is able to stand within minutes and can run after a few hours. It starts to feed itself at three months and is weaned and loses its spots in September or October.

Hunted for centuries

Today the wapiti's once numerous predators are themselves reduced in numbers, due to hunting. They include the wolf, puma, coyote, lynx, wolverine, tiger (in Siberia) and bear, which preys especially on the calves. Adults can use speed to escape or can turn and defend themselves, striking down with the front hooves. A full-grown wapiti stag is credited with being able to break the back of a wolf with one kick.

Contemporary accounts tell how Native Americans sometimes hunted the wapiti in parties, forming a wide crescent around a stag, with the horns of the crescent half a mile apart. As they closed in, the stag dashed first one way and then the other until, when the circle closed, the exhausted stag could be taken alive. Nevertheless, unless the animal was completely worn out, he would stand at bay, and usually one or more of the hunters was injured before the stag was secured. This so-called dance of the deer was conducted for sport. The stag was also killed by Native Americans for his upper canine teeth, which were worn as charms.

Incompatible neighbors

Most of the surviving wapiti are now in national parks or other wildlife refuges, mainly in the western states. Two herds were introduced into the Virginian Jefferson National Forest in the eastern United States, and there have been reintroductions elsewhere in the world, such as New Zealand, Australia and Argentina. Where they invade arable or other settled land, wapiti tend to damage crops or compete with livestock for browse and grazing. They also scrape bark from trees, especially the aspen; wapiti released in Australia destroyed the pokaka. They present the same problem as the red deer in wildlife refuges in Britain; their natural ability to build up numbers under protection is apt to lead to the destruction of their habitat and the need to control them.

WARBLER

THERE ARE ALMOST 300 WARBLERS, named after the melodious song that is characteristic of many species. They are small birds, generally 4–7 inches (10–17.5 cm) long, with fine-pointed bills, long nostrils with a covering flap, medium, rounded wings and weak legs and feet. Warblers are usually drably colored, green, brown or gray. Identification is therefore not an easy task, and many warblers can be identified, even by experienced bird-watchers, only when held in the hand, and even then identification can be difficult. Many warblers migrate, and because they get blown off course, some species turn up in unexpected places. The appearance of such rarities, together with the difficulties of identification, make the warblers particularly interesting to ornithologists.

The warblers are sometimes called the Old World warblers to distinguish them from the New World warblers, also known as the wood-warblers or American warblers. The two groups of warblers are placed in separate families, the Old World warblers, family Sylviidae, having 10 primary flight feathers on each wing and the wood-warblers, family Parulidae, having only 9 primary flight feathers.

The family Sylviidae is sometimes divided into the following subfamilies of Old World warblers: the true warblers (Sylviinae), the gnat-catchers (Polioptilinae), the kinglets (Regulinae) and the wren-warblers (Malurinae).

The Old World warblers are often uniform in form and color, although some tropical warblers are brightly colored. The sexes usually have similar plumage but they differ in a few cases. For example, the blackcap, *Sylvia atricapilla*, is an appropriate name only for the male, which has a black crown contrasting with the pale gray of the rest of his plumage. The female has a brown cap. Similarly, the Rüppell's warbler has a black head and throat with a white mustache. The female retains the mustache but the head and throat are gray and white.

Warblers are restricted to the Old World, except for the Arctic warbler, which has crossed the Bering Straits from Siberia to Alaska, and breeds in western Alaska. The rest are found all over the Old World from western Europe to Australia, with about half in Africa.

Mixed-up migrators

Many warblers spend their lives in one area, whereas others are migratory, especially those that live in places where the winters are cold. For example, only 4 of the 14 species of warblers breeding around Britain spend the winter there. These species are the blackcap, the chiff-chaff, *Phylloscopus collybita*, the Dartford warbler, *S. undata*, and Cetti's warbler, which is a relatively recent colonist.

Some of the migratory species of warblers travel enormous distances on their twice-yearly journeys. The willow warbler, *P. trochilus*, for example, flies up to 7,000 miles (11,200 km), from eastern Siberia to east Africa, and the Arctic warbler may travel even farther, from north-western Europe to Southeast Asia. The Alaskan population of the Arctic warbler returns to the Old World in the winter, crossing the Bering Straits and flying down to Southeast Asia.

Sometimes, reverse migration takes place, and part of a warbler population flies in the opposite direction to the others. This is thought to happen primarily to young birds in their

With almost 300 species of warblers in existence, many with a dull coloration, it is very difficult to identify these birds accurately.

first year. This results, for example, in some warblers arriving in Britain in the fall, by which time the others have left.

Some Siberian breeding warblers arrive in Britain from northern Europe, having flown southwest when they should have flown southeast, notably the yellow-browed warbler and Pallas's warbler. In some years several hundred of each species of warbler can be found in northwest Europe.

Warblers have the typical thin, pointed bills of insect-eaters, and most of them feed entirely on insects, spiders and the eggs and larvae of insects and other invertebrates. They find their food in woods, brushlands and marshes in the bark of trees, in reeds and other plants and

A sedge warbler, Acrocephalus schoenobaenus. Warblers are good singers, though they are not as melodic as the thrushes.

DARTFORD WARBLER

CLASS	**Aves**
ORDER	**Passeriformes**
FAMILY	**Sylviidae**
GENUS AND SPECIES	***Sylvia undata***

WEIGHT
³⁄₁₀–³⁄₁₀ oz. (9–10 g)

LENGTH
Head to tail: 5 in. (12.5 cm)

DISTINCTIVE FEATURES
Long-tailed bird; dark gray upperparts; vinous red underparts; red iris; orange legs; can appear all dark in poor visibility

DIET
Mostly arthropods; occasionally fruit in the fall and winter

BREEDING
Age at first breeding: 1 year; breeding season: March–July; number of eggs: 3 to 5, in 2, sometimes 3, broods; incubation period: 12–14 days; fledging period: 10–14 days; breeding interval: 1 year

LIFE SPAN
2–3 years

HABITAT
Lowland heath, low open scrubland with evergreen shrubs, low trees and aromatic shrubs, and low pine woods

DISTRIBUTION
Parts of southwest Europe, east to Italy, and North Africa

STATUS
Common in good habitat

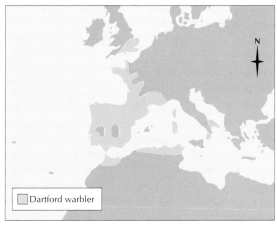

Dartford warbler

sometimes on the ground. Some migrating warblers eat berries, and blackcaps feed at bird-feeders during the winter.

Varied songs

Many warblers advertise their presence by elaborate songs that compete in quality with that of the European blackbird, whereas others have dull songs, like that of the grasshopper warbler, *Locustella naevia*, which can be easily mistaken for the stridulation of a grasshopper.

The breeding habits of warblers are fairly uniform. Each pair usually holds a territory and nests in isolation, although reed and sedge warblers build nests quite close together. The nests are built near the ground, among bushes, reeds or grasses, and are built of grass or reeds, which is woven to form a bowl, ball or a bottle-shaped nest. The males sometimes build several cock nests. The white, spotted eggs, 3 to 10 in number, are incubated by the female alone or by both sexes; both parents usually feed the chicks.

Cryptic species

Some species of warblers are so similar that it is almost impossible to distinguish them by their plumage. Such species are called sibling or cryptic species. The reed warbler and Blyth's reed warbler, for example, can be separated accurately only by a close examination in the hand. The chiffchaff and willow warbler, among others, are distinguished by their song.

The chiff-chaff's irregular, high-pitched *chiff-chaff-chaff-chiff-chaff* is a common sound in the woods during springtime, whereas the willow warbler has a liquid, descending-cadence song, repeated at intervals. The breeding range of these two species of warblers overlaps in most of Europe, but where the willow warbler does not occur, for example in Spain and Portugal, the chiffchaff resembles the willow warbler more closely than it does elsewhere.

In fact, this species of chiffchaff is now considered a separate species, called the Iberian chiffchaff, and there is a zone of hybridization between the two species. The Iberian chiffchaff's legs, which usually are noticeably darker than those of the willow warbler, are lighter in the Iberian Peninsula, and its song is made up of a pattern of notes rather than the two notes of other chiff-chaffs. Presumably the chiffchaff and the willow warbler arose from populations of a common ancestor, but became isolated from each other over time.

An olivaceaous warbler, Hippolais pallida, Lesbos, Greece. Some warblers migrate to spend the winters in warmer climates.

WARTHOG

The warthog's main weapons are its sharp lower tusks, although its upper tusks are larger. The thick skin, matted hair and facial warts all help protect the animal in a fight.

THE MALE WARTHOG grows up to 42–60 inches (105–150 cm) long, exclusive of its tail, which grows to between 16–20 inches (40–50 cm). It stands up to 26–34 inches (65–85 cm) at the shoulder and weighs up to 330 pounds (150 kg). The female is somewhat smaller. The skin of both sexes is slate or clay-colored, with a few bristly hairs over the body and a conspicuous mane of long bristles running from the head down the midline of the back. The most striking feature of a warthog, however, is the very long head, armed with tusks and ornamented with excrescences, or warts, from which the animal derives its name. These are strengthened with gristle and are situated on either side of the face. They are prominent only in the males and do not seem to have any function. The small eyes are set well back, just in front of the ears. The neck is stout and thick, the legs are long and the lengthy, thin tail hangs down when the animal moves slowly but is carried stiffly erect, with the tufted tip hanging over, when it is running.

The curled upper pair of tusks are 12 inches (30 cm) or more in length, though one pair of 27 inches (67.5 cm) has been recorded. They are longer than the lower tusks, which bite against only the lower surface of the upper tusks instead of wearing them away at the points. The upper tusks have enamel at the tips and even this is soon worn away, whereas the lower tusks are coated with enamel throughout their length. The warthog uses its upper tusks when digging up tubers or bulbs.

The young warthog has 34 teeth but in the mature adult there may be only eight because the three pairs of lower incisors and one pair of the upper are lost, as well as all the cheek teeth, except for the last pair of molars. Each of this last pair is large and complex, consisting of a number of long narrow cylindrical denticles, packed closely together.

The warthog is found in most open country in Africa, from Ethiopia to Senegal in the north, southward to the Free State.

Living accommodation

Warthogs prefer open thorn bush, thin forest or plains. They feed both by day and by night, traveling about singly, in pairs or in family parties of one or two sows with their offspring. The boars usually are solitary. At night they lie up in a den, which may be a cave, a hollow under rocks or a

WARTHOG

CLASS **Mammalia**

ORDER **Artiodactyla**

FAMILY **Suidae**

GENUS AND SPECIES *Phacochoerus aethiopicus* (some scientists regard northern race as separate species, *P. africanus*)

WEIGHT
132–330 lb. (60–150 kg)

LENGTH
Head and body: 42–60 in. (105–150 cm); tail: 16–20 in. (40–50 cm); shoulder height: 26–34 in. (65–85 cm)

DISTINCTIVE FEATURES
Bristly pelage, forming mane from forehead to mid-back; gray, brown or black color; white whiskers on lower jaw; 3 warts on each side of head; upward-curving tusks on either side of face

DIET
Grass, fallen fruit, bark, bulbs, tubers; carrion, small mammals

BREEDING
Age at first breeding: 18–20 months; breeding season: year-round; number of young: 2 to 4; gestation period: 170–175 days; breeding interval: 20 months

LIFE SPAN
18 years

HABITAT
Treeless plains and scrub savanna; avoids steep slopes

DISTRIBUTION
Ghana east to Somalia; south to Natal

STATUS
Generally common. *P. aethiopicus delamerei*: vulnerable. *P. africanus aeliani*: endangered

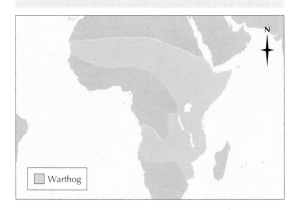

Warthog

depression that they have dug in the ground under the shelter of a dense thicket. They wallow in mud, caking their bodies with it, but their dens are kept scrupulously clean. They sometimes use abandoned aardvark burrows for sleeping or as temporary retreats when disturbed. They enter the burrow backward, presenting their formidable tusks to the entrance, ready to inflict severe wounds on any intruder.

A warthog sow with her piglet. Female warthogs are particularly fearless in defense of their young, and will attack predators as formidable as leopards or elephants if they feel threatened.

Selective grazer

The warthog is principally a grazer, favoring the tender growing tips of short grass. The small incisors are used as tweezers to pluck out the selected food. The warthog's neck is too short for comfortable feeding, so it goes down on its knees to graze, sometimes shuffling forward in this position. Where water is not available or grass is scarce, the warthog may dig up roots, and it is able to survive without water for long periods if necessary. It also takes fruit and berries and, very occasionally, animal food. During a drought that occurred in the Nairobi National Park in 1961, warthogs frequently were seen feeding on the carcasses of wildebeest and other animals that had died of hunger or thirst.

The warthog usually is silent but is prone to grunting when it feeds. It has a good sense of smell and acute hearing but poor eyesight.

Rationed piglets

Usually two to four young are born from October to November after a gestation of 170–175 days. The warthog sow has only four teats. However, in the 20th century scientists discovered that there may be six piglets to a litter, although such a number is unusual, prompting the question of how feeding sessions for more than four young are carried out. Sows seen suckling their young have been standing, although they may lie down in the burrow to do so when the piglets are first born. The young warthogs are a reddish brown color, though they may sometimes be striped.

One warthog lived in the London Zoo for 12½ years. However, there is a record of another specimen living for 18 years in the wild.

Hunted by humans

The lion is the warthog's chief predator in the wild but leopards, cheetahs and wild dogs may take young ones. Several observers have testified that the warthog is virtually fearless. Warthogs have turned and charged pursuing leopards or elephants, which fled at the attack. The sow is particularly aggressive in defense of her young.

Local people hunt warthogs because the flesh is considered to be tasty. Today they are also hunted for sport with jeeps, against which their standard defense methods and maximum speed of 30 miles per hour (48 km/h) are of little use.

Vestigial and rudimentary parts

In general, each part of an animal's body contributes in some positive way to the welfare of the whole. In most animals there are one or more parts that are known as vestigial; they formerly were larger and more vigorous or contributed more to the general economy of the body but have become smaller over time and are, so to speak, wasting away. Scientists refer to other parts of an animal as rudimentary. These are at their beginning and have not yet developed to the point at which they are fully functional. Moreover, not infrequently, animal parts, structures and tricks of behavior occur that seem to be neither functional, vestigial nor rudimentary. Until scientists have discovered more about them, they must be regarded simply as oddities.

The warthog is one of the most unusual of the quadrupeds. Its flat, almost shovel-shaped head, combined with the stout, thick neck, may be useful now, or it may have been useful in the past in perhaps a different kind of habitat, for rooting in the earth or for turning over logs or stones. The warts possibly may serve for additional strength in such exercises, although they may also protect the roots of the teeth. The warts may serve as weapons additional to the tusks for powerful thrusts at an opponent. It is possible that the eyes are set far back because such a position helps to keep them out of harm's way when the animal is rooting for food. If scientists were better informed about the habits and behavior of the animal, it may be that the warthog's distinctively shaped head will prove to be the result of a combination of remarkable adaptations to a specialized way of life.

Warthogs regularly wallow, during which they cake themselves in mud as a means of cooling off.

WATER BEETLE

Fresh water would not remain fresh for long if it were not for scavengers, such as the water beetles, that feed on decaying vegetation. There are several thousand species of water beetles, represented worldwide but most numerous in the Tropics. The name is not wholly appropriate, because not all species are aquatic; some live in damp places among vegetable rubbish, and others in dung. Some of the most familiar water beetles are the diving beetles, family Dytiscidae. These are distinguished from other species in the way that they swim and breathe, and are covered in a separate article.

One of the largest water beetles in the Old World is the silver water beetle, *Hydrophilus* (or *Hydrous*) *piceus*. It is nearly 2 inches (5 cm) long, with smooth, black wing cases. The antennae, usually hidden under the head, are short and club-tipped and aid breathing. There are long labial palps on the mouthparts that resemble a second pair of antennae, and in many species they function as such. A common species in North American ponds and ditches is the giant water scavenger beetle, *Hydrophilis triangularis*. Its body grows to a length of 1½ inches (38 mm). The legs are reddish-black, and the glossy black or brown wing cases often have an olive sheen in strong light. Like *H. piceus*, this beetle has long labial palps and club-tipped antennae.

Awkward swimmers

Water beetles live in shallow, weedy ponds, in pools and in marshes. Some live mainly on damp land; a few live in brackish water or in running water where there are plenty of algae. They swim awkwardly, with alternate strokes of the legs. The first pair of legs is not adapted to swimming, and the beetle uses first the middle and hind legs of one side, then those of the other. These are flattened and fringed.

When the silver water beetle submerges, its underparts appear to be bright and shiny, on account of a covering of very fine, short hairs that trap a thin layer of air. The beetle also carries a bubble of air between the body and the wing cases. Oxygen from the water dissolves directly into the air bubbles on the beetle, acting almost like a natural aqualung. Eventually, however, the beetle must surface to replenish its store of air. It rises headfirst, turning to one side and piercing the surface film with one of the antennae. This forms a funnel that puts the outside air in continuity with the two stores of air the beetle carries. This species, like other members of the family, may fly at night and is attracted to artificial light.

The adult beetle feeds on water plants, including algae, or on decaying matter, seldom on living animal prey, although the larvae are more often predatory. Where larvae live in dung,

A male silver water beetle, Hydrophilus piceus. *The silvery sheen on its underside is created by an air layer that enables the beetle to breathe.*

Water beetles kick their well-developed hind legs to swim or clamber among the aquatic plant matter or rotting debris on which they feed.

WATER BEETLES

PHYLUM	**Arthropoda**
CLASS	**Insecta**
ORDER	**Coleoptera**
FAMILY	**Gyrinidae, Haliplidae, Hydrophilidae, Hygrobiidae, Noteridae**

ALTERNATIVE NAME
Whirligigs (Gyrinidae only); water scavenger beetles

LENGTH
Most species ⅕–⅘ in. (5–20 mm)

DISTINCTIVE FEATURES
Wing cases usually black or dark brown; underside finely haired to trap air, giving silvery appearance underwater

DIET
Adult: decaying vegetation. Larvae: snails, tadpoles, rotifers.

BREEDING
Eggs laid on waterside vegetation or on water in silken cocoons; larvae develop in water, before leaving pond to pupate in soil

LIFE SPAN
From a few weeks to 1 year

HABITAT
Shallow, still or sluggish water, rich in vegetation

DISTRIBUTION
Worldwide apart from polar regions

STATUS
Many species in decline due to habitat loss

---

they are maggotlike, preying on fly maggots. In some parts of the Far East, water beetles are deployed to combat the larvae of other beetles that damage sugarcane and banana stems.

Water babies

The female silver water beetle spins a large, silken cocoon and attaches it to the underside of a leaf among floating vegetation. A vertical chimney of silk projects above the surface, permitting air to reach the 50 to 100 eggs laid inside. Sometimes the cocoons are spun independently of any support and they float like small brown balloons at the surface, with the chimney resembling a mast.

In a few species, the female carries the cocoon with her, holding it between her hind legs. When these hatch, the larvae swallow some air and then bite their way out of the cocoon. The object of swallowing the air seems to be to make the larvae buoyant, so they can rise to the surface to breathe. The carnivorous larvae target mainly water snails. The jaws are asymmetrical and ideally formed for grasping and cutting through snail shells. Well-grown larvae also eat tadpoles. The silver diving beetle larva is nearly 3 inches (7.5 cm) long when fully grown, at which point it leaves the water to excavate a small chamber in damp pondside soil. There it pupates into an adult, which then returns to the pond.

An aquarium rarity

Water beetles are popular additions to aquariums, but the decline of *H. piceus* in Britain, where it has long been rare, serves as an example of how these insects are now under threat. At one time it was believed that overcollecting for aquariums was putting the beetle at risk. This may no longer be the case, but the species is now in grave danger due to agricultural operations that involve filling in ponds and the dredging of drainage ditches in former marshland. In addition, fertilizer runoff causes algal blooms in fresh water, rendering it unfit for animal life. It has been estimated that, since 1970, the silver water beetle has suffered a decline of up to 50 percent in Britain, where its combined locations add up to an area of only 60 square miles (150 sq km).

WATERBUCK

THE WATERBUCK IS A large antelope that is always found near rivers but which, in spite of its name, lives on drier ground than the closely related kob (*Kobus kob*) and lechwe (*K. leche*). It stands 44–52 inches (1.1–1.3 cm) at the shoulder and may weigh 450–500 pounds (200–230 kg). It has a coarse, rough coat that tends to be darker and grayer in the male and redder in the female. There is white fur on the midline of the belly, around the muzzle and around the eyes. There also is an indistinct white band around the neck behind the ears, and either a white ring around the rump or a white patch in that region. The feet are blackish. The neck and haunches of adult males are thick, and some have a shaggy beard, both features that offer protection from horn thrusts. The coat is also very greasy and smelly, possibly to aid in scent communication or to serve as waterproofing.

The males have horns that are long, slender and curved. These are ringed for much of their length with annuli, growth rings that give some indication of a male's age. Males use their horns for sparring over territorial or mating rights.

Separate species

Waterbuck occur over much of sub-Saharan Africa. Those subspecies with a white rump patch are known as defassa or singsing and are found in the more westerly parts of the range. Those with a white ring on the rump are known as common waterbuck and are found more to the east. The two were once thought to be different species, but biologists now know that they interbreed where their ranges meet. Thus, in parts of East Africa such as the Nairobi National Park, the herds cannot be definitely classed as one species or the other, but show all intermediates between the two. South of the East African parts of their range they are separated by geographic barriers. In Zambia, their ranges are divided by the Muchinga escarpment, and farther south by desert. Moreover, each type shows geographic variation, and there are altogether about seven subspecies of waterbuck: three of the common and four of the defassa, differing in color, the amount of white around the eye, length of ear and other details.

Typically found at the boundaries of grassland and shrub or woodland, waterbuck venture into arid country, but they always stay within reach of water. They occur on the savanna of southern Gabon, which they have colonized by crossing the Zaire River from the south, rather than by penetrating the forest belt from the north. In arid parts of Somalia, they are restricted to the major river valleys. The thick cover by the rivers gives them protection. They come out to graze in the grassland beyond, returning again to rest.

Waterbuck are naturally day-active; where they have not been hunted, as in the Nairobi Park, they stay in the grasslands all day and return to cover at dusk. Where they have been constantly shot at and disturbed, as in Zaire and Somalia, they emerge at night and lie up during the day. This is an example of how an animal's habits may change under human pressure.

Territorial males

An adult male waterbuck holds a territory of ¼–1 square mile (0.7–2.6 sq km), extending along the river and back into the grazing areas. He moves about daily to and from the riverine thickets within the territory, the boundaries of which are clearly defined by the owner's behavior, at least along the river, although less clearly farther inland. Scent is not important in marking the territory, urine and dung being voided more or less at random. The length of its river frontage is

A male waterbuck in Matusadona National Park, Zimbabwe. This antelope readily wades up to its belly to reach fresh plant matter, and sometimes enters water when fleeing predators.

A defassa waterbuck pauses to sample the air in the Masai Mara, East Africa. Scent and hearing are vital to waterbuck, which are preyed on by lions, leopards and hyenas.

COMMON WATERBUCK

CLASS	**Mammalia**
ORDER	**Artiodactyla**
FAMILY	**Bovidae**

GENUS AND SPECIES **Common waterbuck, *Kobus ellipsiprymnus***

WEIGHT
450–500 lb. (200–230 kg)

LENGTH
Head and body: 75–88 in. (1.9–2.2m); shoulder height: 44–52 in. (1.1–1.3m); tail: 14–18 in. (35–45 cm)

DISTINCTIVE FEATURES
Large antelope; gray-brown to rufous-yellow upper body, paler below; greasy, pungent coat; white ring or patch on rump; male has horns, each with 18 to 38 annuli

DIET
70–90 percent grass; some foliage

BREEDING
Age at first breeding: female 13 months, male 14 months; breeding season: year-round; gestation period: 227–287 days; number of young: 1, rarely 2; breeding interval: 2–5 weeks after giving birth

LIFE SPAN
18 years in captivity

HABITAT
Savanna grassland with gallery forest and woodland patches; close to water

DISTRIBUTION
Sub-Saharan Africa; defassa found from extreme west across to East Africa and south to Angola; common waterbuck found primarily to east

STATUS
Low risk

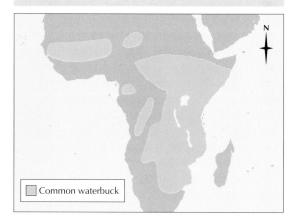

Common waterbuck

an indication of the rank of the male holding it. Older territorial males tend to recognize one another's boundaries rather than try to invade and usurp. Often, mere posturing, side on to the neighbor with neck arched and tail extended, is enough to reestablish the ground rules. Younger males, by contrast, are much more ready to pick a fight. The aggressors horn the ground and circle one another before engaging in a head-to-head butting contest, each trying to thrust his horns at the other's flanks. These fights can end in death, or at least a bad gash or broken horn.

Looking for mates
Each male tries to attract females into his territory. If a group of females wanders through it, he tries to keep them, while his neighbors

move up to the boundary, ready to herd the females the moment they move out. The male herds the females by running in front of them and standing with his head up and forefeet together, blocking their path, and while they are moving he gives chase, running alongside and butting their flanks.

The females have their own home ranges, which are used for grazing. They are similar in extent to a male's territory but are not defended. A group of females may therefore move unchallenged into a neighboring group's home range and out again as occasion demands. The home ranges, moreover, cut across the males' territories, so the females are constantly harassed as they wander to graze. They spend the night with their young in the riverside bush in groups of three or four. During the day, they move about the home ranges in larger groups of up to 30. A few subadult males move along with the females, but those past puberty are usually evicted by the territory owners, and they join a peripheral bachelor herd until each can claim his own territory.

Lick and kick

When a male greets a female, he investigates her to find out if she is in estrus. He sniffs and licks her vulva and tail, often letting her urine run over his muzzle and lips. He may adopt the *Flehmen* pose (head raised, lips drawn back, nose wrinkled) to inhale and assess her scent. He then performs the *Laufschlag*, putting his foreleg between the female's legs or under her belly. He mounts the female without necessarily copulating. If she is not receptive, the female moves away, while nodding her head to appease him.

A male grazes alongside females he has lured into his territory. A female in estrus travels a greater distance and grazes for a longer time each day than one who is not. She thus passes through a greater number of male territories and sexual communication is more complex than usual. The male may, for example, rub against the female with his face and the base of his horns, where there is a depression that is richly supplied with nerves. He often puts his horns on either side of her rump and pushes. For her part, a receptive female often shows excitement, nuzzling a male's groin or even mounting him.

Concealed calf

Gestation takes 227–287 days, and the single calf, rarely two, is born in riverside cover. In the early days, the mother spends longer than usual in the thicket, leaving her young for as short a time as possible and nursing it three or four times in a 24-hour period. After 3–4 weeks, the calf begins to come into the open. It joins a nursery group with other mothers and calves, and indulges in play. After it is weaned at 6–8 months, the calf wanders away from its mother to take its place in a bachelor or spinster group. Although the sex ratio is thought to be about equal at birth, only 30 percent of the adult population are males.

On the savanna of Kwazulu-Natal, South Africa. The waterbuck is a normally sedentary species, though it tends to wander during the rainy season.

WATER-HOLDING FROG

Water-holding frogs have evolved a number of adaptations to life in an arid environment, including a bladder that can hold up to half of their body weight in water.

ANY FROG OR TOAD IS LIKELY to discharge a fluid, mainly water, from its bladder if it is gently squeezed. Frogs and toads living in deserts hold more water in their bladders and are therefore termed water-carrying or water-holding frogs. Naturalists have singled out one in particular for this name: the water-holding frog of the deserts of Australia.

This frog is squat-bodied, 2¾ inches (7 cm) long, with fairly short, stout front legs and plump hind legs with webbed toes. It is greenish gray, often with a dark line down the middle of the back, and its skin is warty. The head seems small for the bloated body; the animal looks more like the European common toad than the common frog. This may be why this and other similar species in Australia have at times been called water-holding or water-carrying toads. The range of the water-holding frog covers the dry areas of New South Wales and Queensland, northern South Australia and central Australia.

Tanker frogs

Frogs and toads can lose water rapidly through their skin, but they also can rapidly absorb water through it, which is why they never need to drink. Most of this water is stored in the lymph spaces under the skin and between the muscles. Much of it is stored in the bladder and can be taken back into the rest of the body if needed. This can best be appreciated by comparing the water-holding frog with the African clawed frog, *Xenopus laevis* (discussed elsewhere). The clawed

frog continually lives in water and has a small bladder that holds only 1 percent of its body weight. The bladder of the water-holding frog can hold up to 50 percent of its body weight of water.

Cocooned for the dry spell

More than half the surface of the Australian continent has 10 inches (25 cm) or less of rain each year. The driest places are in the heart of the continent, where the annual rainfall is less than 5 inches (12.5 cm) a year. In this area, the rain falls as heavy but infrequent showers, and in a given district there may be a lapse of several years between one rainstorm and the next. During periods of drought the water-holding frog lies buried in the ground, where the soil is permanently moist. It also sheds a cocoonlike bag of skin around itself by casting off a cell-deep outer layer of its skin. This encases it completely except at the nostrils, and water accumulates between the skin and the skin-bag, inhibiting evaporation. The frog remains in a torpid state, eating nothing, until the next rainstorm. Then it casts off the skin-bag and comes to the surface to feed rapidly on insects that become abundant during the wet spell. At the same time the frog replenishes its water supply and breeds.

Brief infancy

Water-holding frogs breed in temporary pools caused by heavy rains, in the same way as pond-breeding frogs and toads everywhere. The main difference is that the eggs hatch in much less than the usual time and the tadpoles develop much more quickly. They become froglets in less than 2 weeks, compared with, for example, the 10 weeks taken by the European common frog, *Rana temporaria*. The young frogs feed heavily and rapidly and also fill up with water; then they bury themselves in the ground before the hot sun dispels the moisture remaining from the rainfall.

It is hardly surprising that little precise information is available on many aspects of the biology of animals living under such rigorous conditions. It is only within recent years that a few of the animals in deserts near the more densely populated regions of the world, where biologists are more numerous, have been subjected to close scrutiny. It seems reasonable to suppose, however, that the main hazards in the life of water-holding frogs are connected with the very brief period of infancy. When so much development, feeding and water storage must be carried through in so short a space of time, there must be many tadpoles and froglets that fail to survive. Predation, by birds and reptiles principally, while the frogs are above ground, also regulates population numbers. Populations are large mainly because of explosive breeding strategies and good environmental conditions

Aboriginal food and drink

Some water-holding frogs and toads occur outside Australia, but these have received less scientific attention than their Australian relatives. This is partly because their adaptations are less

extreme, but it is due primarily to the link between the amphibians and the aborigines of Australia. For many years, tradition held that when aborigines felt thirsty they might dig a water-holding amphibian out of the ground, hold it over their open mouth and squeeze it to quench their thirst. This is, it seems, something of an oversimplification. In his *Book of Australian Wild Life*, Australian naturalist Harry Frauca reveals that water-holding frogs are "said to be found in large numbers in the claypans of Centralia and to provide some aborigines with food and drink. There are reports to the effect that some tribes will dig up water-holding frogs from the claypans, squash them and drink the liquid contained inside the bodies and later on throw the frogs into the cookpot."

Amphibians redefined

Frogs and toads belong to the class Amphibia, the name meaning loosely "animals leading double lives." It is a common error to presume that amphibians spend half their time in water and half on land. In fact, some, such as the clawed toad, spend all their time in water. There are many that never go into water, including the Stephens Island frog, *Leiopelma hamiltoni*, of New Zealand, 80 different kinds in New Guinea and adjacent islands and four in Australia. All of them, when not in water, live in moist places, where evaporation from the skin can be counterbalanced by intake of water through the skin, or else must take their water supply with them. Amphibians are therefore not so much animals spending half their time in water and half out, but animals adapted to ensuring a supply of water whether they are immersed in it or outside it. Out of water, however, desert-living amphibians can lose the equivalent of a third of their body weight in water without dying.

During periods of drought, the water-holding frog buries itself as far as 3 feet (1 m) down in the earth in order to help keep its skin moist.

WATER-HOLDING FROGS

CLASS	**Amphibia**
ORDER	**Anura**
FAMILY	**Hylidae**
GENUS AND SPECIES	***Cyclorana platycephala*** (detailed below); others

LENGTH
2¾–4 in. (7–10 cm)

DISTINCTIVE FEATURES
Squat, squarish body; warty skin; greenish gray color, often with dark line down midback; head small relative to body size; short, stout forelegs; plump hind legs, webbed toes

DIET
Insects

BREEDING
Breeds only above specific temperature. Breeding season: after spring or summer rains. Tadpoles metamorphose within 2 weeks in ideal conditions.

LIFE SPAN
Not known

HABITAT
Arid areas

DISTRIBUTION
Australia

STATUS
Common

Water-holding frogs

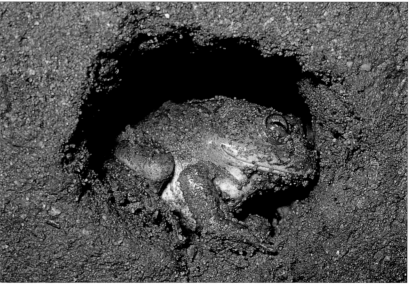

WATER SCORPION

The water stick insect, Ranatra linearis. Its abdomen extends into a whiplike structure that is used like a snorkel to take air from above the water.

WATER SCORPIONS ARE BUGS and are thus entirely unrelated to land scorpions (class Arachnida); they are also far less dangerous to humans. These aquatic insects are called scorpions because of the shape of their front legs, which are modified for grasping prey, and because of their long, slender tail, which resembles, to a slight degree, a scorpion's tail. The larger kinds of water scorpions can pierce human skin with their beaks, with painful but not serious consequences. In Australia, people are bitten often enough for the insects to have earned the name toe-biters or needle bugs.

Snorkeling

Water scorpions are flat and blackish-brown, the largest being no more than 1–2 inches (2.5–5 cm) long. They live on the bottom around the edges of sluggish waterways, ponds and ditches. The roughly oval outline of the body appears leaflike in some species because the drawn-out abdomen resembles a stalk. Poor swimmers, water scorpions seldom venture into open water, but may climb about on waterweeds to get to the surface and breathe air. For this they use the abdominal process, which consists of two half-tubes closely applied to each other and held together by interlocking bristles to form a breathing siphon, the tip of which is pushed above the water surface.

The family that includes the water scorpions is small, with only about 150 species, but these are well distributed over the world. The water stick insect, genus *Ranatra*, is in the same family.

Evolved for the underwater life

When the wing cases of a water scorpion are raised, a pair of delicate hind wings are revealed. These are seldom used, and in some species the principal wing muscles are so reduced that flight is impossible. The wings are pink with bright red veins, and the part of the hind body over which they lie is brick red with black bands. Also on the abdomen are three pairs of false spiracles. The spiracles of insects are primarily breathing pores, but in the water scorpion a number of spiracles have evolved other purposes. Some have become balancing organs. Experiments have shown that these spiracles are extremely sensitive to water pressure, which will be very slightly greater on whichever flank of the insect is tilted downward.

Grab and stab

Water scorpions are predatory bugs. They lurk among the aquatic debris and plant matter, ambushing small tadpoles, insects and other animals. They seize prey with their highly modified forelegs. Each of these is hinged like a jackknife to close on a victim. The end portion of each leg folds back into a groove along the basal part when not in use. The piercing beak, or rostrum, a characteristic of the bug order, Hemiptera, works like a hypodermic needle both to inject and to extract. The water scorpion plunges it into the victim, then pumps the prey full of digestive juices. The juices dissolve the body tissues, which are then sucked back in liquid form. As well as being digestive, the injected juice is venomous and rapidly paralyzes and kills the prey. Even large victims, such as young fish, are quickly subdued.

Chains of eggs

Water scorpions lay their eggs in spring among aquatic plants near the water surface. At one end of each egg is a bunch of seven long filaments. These become entangled and the eggs cling to each other in chains. The newly hatched young resemble their parents except in size and the fact that they lack wings.

Stick in the water

Water stick insects are larger than most water scorpions. They are long and slender with long legs, rather like the familiar stick insects on land for which they are named. The breathing siphon is nearly as long again as the rest of the body, which is a pale ocher color, and the method of feeding is similar to that of the water scorpion. Water stick insects live in still waters among waterweeds and standing reeds. They swim by waving their forelegs and kicking with the middle and hind pairs. They can also climb. Unlike water scorpions, water stick insects have functional wings and readily fly in search of a new habitat if their pond or ditch dries up.

Instead of casting her eggs loosely among plants, the female water stick insect inserts them within the stems of floating leaves, such as those of water lilies. She does this by using her sawlike ovipositor (egg-laying organ) to cut a slit in a stem for each egg.

WATER SCORPIONS

PHYLUM	**Arthropoda**
CLASS	**Insecta**
ORDER	**Hemiptera**
FAMILY	**Nepidae**

GENUS AND SPECIES **Several, including water scorpion, *Nepa cinerea*, and water stick insect, *Ranatra linearis***

ALTERNATIVE NAMES
Toe-biters; needle bugs (water scorpions)

LENGTH
⅗–2 in. (1.5–5 cm)

DISTINCTIVE FEATURES
True bugs with piercing mouthparts; front legs adapted to grabbing and holding onto prey; long siphon, or breathing tube, at hind end of body used to conduct air from above water to body beneath

DIET
Smaller arthropods; small fish, tadpoles

BREEDING
Water scorpion lays eggs in long strands; water stick insect cuts slit in host plant and inserts eggs

LIFE SPAN
Not known

HABITAT
Slow-moving or still bodies of fresh water, often among aquatic vegetation

DISTRIBUTION
Family found worldwide

STATUS
Not threatened

A water scorpion, Nepa cinerea, *devours its bloodworm prey in a pond in England. This species is restricted to the Old World.*

WATER SHREW

SUPERB SWIMMER and diver, the water shrew resembles nothing so much as an animated bubble in the water because of the air trapped in its fur, which gives it a silvery sheen. A common species in North American habitats is *Sorex palustris*. This species is 5–6 inches (12.5–15 cm) long from head to rump, with a tail up to 3½ inches (9 cm) long. Its body looks bulky, but an adult weighs no more than ⅔ ounce (18 g), even less in the winter. The upperparts vary from slate gray to dark brown, and the silvery white of the underparts, cleanly separated from the upper, makes a stark contrast. The water shrew's snout is short and broad by shrew standards, and its eyes are small and blue. The ears, each bearing a tuft of white hairs, are entirely hidden beneath the fur. The feet are brown and broad, the hind feet being the longer. The toes are bordered with stiff, bristly hairs, which make them efficient paddles. They also are used in grooming the fur. The tapering tail of the adult, flattened from side to side, has a double fringe of strong, silver-gray hairs along its underside, constituting a keel and making the guiding tail more efficient as a rudder in the water.

S. palustris ranges from Alaska to the southernmost extent of the Rockies and east to Nova Scotia; an isolated population lives in the Appalachians. The Pacific water shrew, *S. bendirei*, ranges from British Columbia to northwest California.

The American water shrew's counterpart over much of Europe, as well as southwestern and northern Asia, is the European water shrew, *Neomys fodiens*. In southern Europe this species is replaced by the Mediterranean water shrew, *N. anomalus*, which is a little smaller and has upperparts of a slightly paler brown; it also lacks the keel of hairs on the underside of the tail.

Watery lifestyle

Water shrews live on the banks of streams or lakes. They are active throughout the day, with alternate periods of feeding and rest, though activity peaks before sunrise and after sunset. They seldom venture more than a couple of yards from the bank, swimming buoyantly with the head slightly above the surface and the body flattened. Water shrews appear to be able to walk for a time along the bottom of a stream, and the bristles on their feet enable them to scull over the water's surface for brief periods, perhaps to chase insects. Their range of vision is limited, so water shrews are alarmed by sudden noises.

A water shrew makes a shallow burrow in a bank for its sleeping quarters, but does not hibernate: in winter it may even be seen chasing prey beneath the ice. It gives a cricketlike chirp, near the upper limit of human hearing and audible to younger persons but less so to the elderly, as with the squeaks of some bats.

A venomous bite

Whirligig beetles and water gnats on the surface are prey for water shrews, and at the bottom these busy mammals forage for insect larvae. They eat a variety of other aquatic animals such as snails, worms, small crustaceans, frogs and small fish. They are not averse to carrion. Their bite is venomous, and tests have shown that a secretion from their submaxillary glands can be lethal even to small rodents.

Short life span

During the December–September breeding season, the female makes a new, deep burrow in the bank, ending in a small nursery chamber lined with moss and fine roots, or in a round nest of woven grass or leaves. After a gestation period of 20 days up to eight blind, naked young are born, each weighing just over ⅓₀ ounce (1 g). They develop rapidly, are weaned in 27 days and are independent at 5–6 weeks old. There may be two or three litters per year. The life span is 15 months or more, but it is common for water shrews to die within the year of their birth.

A European water shrew on the riverbank with its crane-fly prey. Though classified as insectivores, shrews are fiercely predatory and will eat almost anything they can overpower.

AMERICAN WATER SHREW

CLASS	**Mammalia**
ORDER	**Insectivora**
FAMILY	**Soricidae**
GENUS AND SPECIES	***Sorex palustris***

WEIGHT
⅓–⅔ oz. (10–18 g)

LENGTH
Head and body: 5–6 in. (12.5–17 cm);
tail: 2¾–3½ in. (6–9 cm)

DISTINCTIVE FEATURES
Long, narrow snout; small eyes and ears; fur usually black or gray-black on back, silvery white on belly; long tail; long, stiff bristles on feet

DIET
Aquatic insects, including larvae and nymphs of caddis flies, crane flies, mayflies and stoneflies; flies, earthworms, snails; occasionally small fish; fungi, plant matter

BREEDING
Age at first breeding: 6–8 months; breeding season: December–September; gestation period: 20 days; number of young: 3 to 8; breeding interval: 2–3 months

LIFE SPAN
15–18 months maximum, usually shorter

HABITAT
Burrows in the bank of streams, marshes, rivers, bogs; favors northern forests and streams with plenty of cover, including logs and thick vegetation

DISTRIBUTION
Southern Alaska south to southern Rocky Mountains, across Canada to Nova Scotia in east; also in Appalachians

STATUS
Not threatened

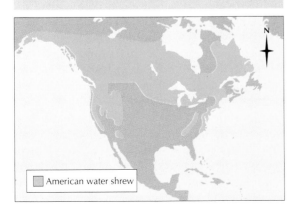

American water shrew

Elemental threats

Shrews are unpalatable to most mammalian predators. A few mammals and birds are resistant to, or at least tolerant of, the shrews' noxious chemical defenses. In Britain, water shrews constitute up to 6 percent of all shrews found in the owl pellets of tawny and barn owls. Shock is usually cited as the primary cause of death in shrews captured for field research.

Even in the absence of natural or human enemies, a water shrew has such a rapid metabolism that it must eat every 2–3 hours or risk dying of starvation. The water shrew is even worse off than other shrews in that it cannot remain long in water because it cannot stand long periods of exposure to cold. As soon as it has grabbed some food in the water, it emerges onto the bank. It runs through its tight tunnel to squeeze the water from its fur, and then uses its fringed hind feet to comb remaining moisture from the fur. This is essential to prevent death from exposure. Not only does a water shrew eat its own weight of food in 24 hours, it is almost certainly unique in that it continues to eat while grooming its fur. Most animals have a meal and then indulge in a leisurely grooming session, but in the hectic life of a water shrew there is no time to waste on such luxuries.

The European water shrew can reach depths of 30 inches (75 cm) or more in still water, although the buoyancy of its air-filled coat prevents it from staying down for long.

WATER SNAKE

One of the three European colubrids is the viperine snake, Natrix maura, *pictured, which sometimes resembles a viper with its zig-zag patterning.*

THE NAME WATER SNAKE IS likely to be given to any snake that spends most of its life in or near water and feeding on aquatic animals. Most water snakes belong to the family Colubridae, but those that have the best claim to the name, the file snakes, are in a family of their own, the Acrochordidae. File snakes live on the coasts from India to the Solomon Islands and are nearly as well adapted to aquatic life as the sea snakes. The largest water snake, *Acrochordus javanicus*, sometimes called the elephant's trunk snake, grows up to 9 feet (2.7 m) in the largest females. The nostrils are on the top of the snout and can be tightly closed by muscular valves. The body is streamlined, as the scales are placed edge to edge rather than overlapping. For this reason, these Far Eastern water snakes are used as snakeskin leather for shoes and handbags.

A very widespread group of water snakes belong to the family Colubridae in the closely related genera *Natrix* and *Nerodia*. These snakes are found in North America, Europe, North Africa and much of Asia, and can be regarded as one of the most abundant and successful of the nonvenomous snakes. There are 11 species in North America. The most widespread of these is the northern water snake, *Nerodia sipedon*. As with so many snakes that have a large geographical range, there is immense variation in color and pattern. One common form has red-brown transverse markings across the body, separated by cream or gray bands that are thinner on the top of the back than they are at the sides.

However, several other species of North American water snakes can also look like this. Where their ranges overlap, water snakes can be very difficult to identify.

There are three species of water snakes in Europe. The grass snake, *Natrix natrix* (discussed elsewhere), is the least aquatic of these. It lives mainly on land, but frequently enters the water to look for the frogs on which it mainly feeds. Viperine snakes, *N. maura*, so-named because they often have a zigzag pattern on the back and a dark V-shaped mark behind the head, and therefore superficially resemble vipers, spend even more time in the water, and fish form a large part of their diet. Dice snakes, *N. tesselata*, which are sometimes called tesselated water snakes, are the most aquatic of all. They spend the majority of their lives in the water, emerging to bask in the sunshine from time to time, and feed exclusively on fish.

Water snakes are found on all the other continents where there are warm or hot climates. Most of them are not closely related to one another. Their common characteristic is that they spend a lot of time in the water, usually hunting there for their prey.

Defensive strikes

Water snakes are either nonvenomous or, as in the rear-fanged water snakes, only mildly so. When disturbed, they escape by diving into water, but if cornered, many of the North American water snakes will turn and attack. The northern water snake, which is usually called the common water snake in the northeastern United States and may grow to 4 feet (1.2 m) long, flattens its body and strikes with its mouth open, sometimes drawing blood. No venom is injected but there is a danger of blood poisoning if the wound is not cleaned. The snake's final line of defense is to emit an evil-smelling liquid. European water snakes are not so aggressive. They are more likely to play dead, but they do produce a similar unpleasant-smelling liquid to ward off potential predators.

The northern water snake lives in marshes, streams and lakes but, like other water snakes, spends much of its time basking, draped over branches that overhang the water. Other water snakes are less attracted to the water and often may be found on land some distance from water.

WATER SNAKES

CLASS	**Reptilia**
ORDER	**Squamata**
SUBORDER	**Serpentes**
FAMILIES	**Acrochordidae, Colubridae**

GENUS AND SPECIES **File snakes: elephant's trunk snake, *Acrochordus javanicus*. North America: green water snake, *Nerodia cyclopion*; red- or yellow-bellied, *N. erythrogaster*; banded, *N. fasciata*; northern, *N. sipedon*. Europe: viperine snake, *Natrix maura*; grass snake, *N. natrix*; dice snake, *N. tesselata*. Others.**

ALTERNATIVE NAME
Common water snakes (many species of North American water snakes)

LENGTH
File snake: up to 9 ft. (2.7 m); green water snake: 74 in. (1.9 m); most adults half this size

DISTINCTIVE FEATURES
Very varied coloration and markings; spend most time in water

DIET
Very varied. Invertebrates including mammals, birds, reptiles, amphibians, fish; crayfish and other invertebrates

BREEDING
North American water snakes and file snakes: viviparous (give birth to live young); European species: egg-laying

LIFE SPAN
Not known

HABITAT
Fresh water, occasionally slightly salty water

DISTRIBUTION
Worldwide in warm or hot climates

STATUS
Many species extremely abundant

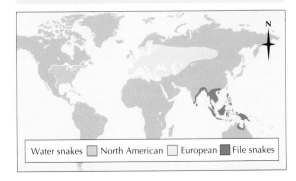

Water snakes ▢ North American ▢ European ▢ File snakes

Bands, blotches and stripes

Some of the North American species, such as the green water snake, *Nerodia cyclopion*, have no distinctive pattern. These snakes are found along the Gulf coast in Texas, the Mississippi valley, Florida and northward to South Carolina. Most species, however, have distinct but immensely variable patterns. For example the banded water snake, *Nerodia fasciata*, usually has dark bands running transversely across the body. The bands may be red, brown or black, and the background color may be gray, brown, yellow or red. In the Mississippi valley the bands on these snakes are broader than on snakes from elsewhere. Along the Gulf coast, from Texas to Florida, most specimens have longitudinal stripes running along the body. Banded snakes from the mangrove swamps of southern Florida may have any of these characteristics. Some authorities regard these forms as subspecies, terming them the Florida, broad-banded, Gulf salt marsh and mangrove water snake, respectively.

Another species is even more variable and confusing. It has no English name that is common to all parts of its range, and its scientific name is *Nerodia erythrogaster*. In different parts of its range, this snake is called a red-bellied water snake, a copperbelly (from Delaware to Florida and eastward to Alabama), a yellow-bellied water snake (from Georgia and Iowa, southward to Texas), a plain-bellied water snake (in New Mexico) or a blotched water snake (from Missouri and Kansas to northeast Mexico).

Red-bellied water snake, Nerodia erythrogaster. Depending on its range, this snake may be called a copperbelly or a yellow-bellied, plain-bellied or blotched water snake.

WATER SPIDER

A water spider traps bubbles of air beneath the abdomen by means of its long hindmost pair of legs, which, like the abdomen, are covered with hairs.

ALTHOUGH MANY SPIDERS are able to live temporarily underwater or will even enter water voluntarily, there is only one species that lives more or less permanently below the water's surface. It does so by constructing the equivalent of a diving bell.

There is nothing particularly distinctive about the appearance of the water spider. It is small-bodied and long-legged, the front part of the body light brown with faint dark markings, the chelicerae (an anterior pair of appendages) reddish brown and the abdomen grayish and covered with rather short hairs. An unusual feature is that the females are usually smaller than the males, the size range being ⅕–⅗ inch (5–15 mm), although females of up to 1 inch (28 mm) long have been recorded. The water spider ranges across temperate Europe and Asia.

Stocking up with air

Although the adult water spider lives permanently in water, it is dependent on air for breathing. It rises to the surface and hangs head-down from the surface film with the end of its abdomen and the tips of the hindmost pair of legs pushed up into the air. With a sudden jerk of the abdomen and the hind pair of legs, a bubble of air is trapped on the spider's underside. The air bubble covers the body and is held in place by the legs, which are swept over the abdomen as the spider dives to trap and hold more air.

The spider then descends, swimming down or climbing down the stems of water plants, to its thimble-shaped bell of silk, holding the bubble of air between its hind legs. It enters and, with its head directed toward the top of the bell, slides the bubble of air forward under its body. The front part of the bubble is then released to rise to the top of the bell. The spider then turns around, directing the tip of the abdomen upward, and releases the rest of the air, stroking the abdomen with the rear legs if necessary to brush it off. The spider then goes to the surface and swims down with another air bubble, this being repeated until the bell is filled with air.

Building the bell

The bell is made by first spinning a platform of silk between the stems or leaves of water plants, with strands running out from the spider to the vegetation around. Wherever the spider goes, it lays down a guideline of silk and this may serve other purposes than guiding the spider back to its home. The thread accumulation probably helps to secure the silken bell, and insects bumping into them probably alert the spider to the approach of prey. Once the platform of silk has been spun, the spider releases air beneath it, making the silk web bulge upward. As more and more air is added, the web takes on the shape of a bell or thimble.

Lying in ambush

During the day the spider remains inside the bell, with the front legs pushed beyond the mouth of the bell into the water. Water spiders are sensitive to vibrations in the water. Any small animal passing nearby, particularly an insect or its larva, will prompt the spider to dash out, seize it and return to the bell to eat it. An insect falling on the surface of the water also will set up vibrations, sending the spider to the surface to seize the insect and take it down to the bell to be eaten. Water spiders may also feed on small fry or tadpoles. By night the spider leaves its bell to hunt, but it always returns to the bell along with its prey.

Aviating submariners

Mating begins in spring or early summer, with the male loading his palps with sperm while still in his own bell and then setting out to find a female in her bell. If she is ready to mate, only a brief courtship ensues; otherwise she lunges at him, making him retreat. Having mated with her, he may remain in her bell for a while, and may even mate a second time. The female lays 50 to 100 eggs in a silken bag that takes up the upper half of the cavity of the bell. Depending on environmental factors such as water temperature, the eggs hatch in 3–4 weeks, the spiderlings biting their way through the bag into the bell, where they stay for a few weeks, molting twice during that time. Some of the brood stay in the same pond, but many rise to the water's surface, climb out, spin threads of silk onto the wind and float away to disperse to new ponds.

Predators preyed on

There probably is a heavy mortality as the young spiders disperse. Even after this stage of the life cycle has been safely passed, predators are numerous. They include dragonfly larvae, back swimmers, water stick insects, beetles and their larvae, frogs, fish and possibly larger members of

their own species. Biologists are still unsure as to whether cannibalism in water spiders occurs in the wild or is the result of being kept in captivity.

As winter approaches, water spiders go to a lower level in the pond and spin an entirely enclosed winter bell, stocking it with air. Some water spiders may use an empty water snail shell lying on the bottom. The one bubble of air will last 4–5 months, because the spider is completely immobile and using almost no energy while it is in its submerged winter quarters.

Submerging spiders

Among other spiders that voluntarily submerge in water, one species, *Dolomedes triton*, has been tested experimentally. Belonging to the family Pisauridae, *D. triton* is unrelated to the water spider. W. H. McAlister, of the University of Texas, found that this spider requires a solid surface to push itself into and out of the surface film, and while submerged needs a solid support to cling to as an anchor. It is scientifically correct, therefore, to say that the spider deliberately enters water. In addition, it was found to remain submerged voluntarily for 4–9 minutes, exceptionally up to 30 minutes. While being tested, the spider was found to survive sustained immersion for up to 180 minutes, which is 10 times as long as most terrestrial spiders survive.

The water spider spends almost all of its time underwater with its air-filled nest, a silk sheet attached to aquatic plants that contains a trapped bubble of air.

WATER SPIDER	
PHYLUM	**Arthropoda**
CLASS	**Arachnida**
ORDER	**Araneae**
FAMILY	**Argyronetidae (or Agelenidae)**
GENUS AND SPECIES	***Argyroneta aquatica***

ALTERNATIVE NAME
European water spider

LENGTH
⅕–⅗ in. (5–15 mm)

DISTINCTIVE FEATURES
Long hairs on two hindmost pairs of legs and short hairs on abdomen to help trap air

DIET
Invertebrates; some small vertebrates, including fish fry and tadpoles

BREEDING
Fertilization in female's bell; egg sac deposited in bell top; spiderlings disperse after several molts

LIFE SPAN
Not known

HABITAT
Slow-flowing pools, ponds or streams

DISTRIBUTION
Europe and Asia

STATUS
Common

Index

Page numbers in *italics* refer to picture captions.
Index entries in **bold** refer to guidepost or biome and habitat articles.

Page numbers in *italics* refer to picture captions. Index entries in **bold** refer to guidepost or biome and habitat articles.

Page numbers in *italics* refer to picture captions. Index entries in **bold** refer to guidepost or biome and habitat articles.